RHS
LATIN
for
GARDENERS

RHS Latin for Gardeners
by Lorraine Harrison

First published in Great Britain in 2012 by Mitchell Beazley,
an imprint of Octopus Publishing Group Ltd,
Endeavour House, 189 Shaftesbury Avenue, London WC2H 8JY
www.octopusbooks.co.uk

An Hachette UK Company
www.hachette.co.uk

Published in association with the Royal Horticultural Society

ISBN: 978 1 84533 731 5

A CIP record of this book is available from the British Library

Set in Garamond Pro

Printed and bound in China

Octopus Publisher: Alison Starling
RHS Publisher: Rae Spencer Jones
RHS Consultant Editor: Simon Maughan

Conceived, designed and produced by
Quid Publishing
Level 4 Sheridan House
Hove BN3 1DD
England

Design: Lindsey Johns

The Royal Horticultural Society is the UK's leading gardening charity dedicated to advancing
horticulture and promoting good gardening. Its charitable work includes providing expert advice and information,
training the next generation of gardeners, creating hands-on opportunities for children to grow plants and
conducting research into plants, pests and environmental issues affecting gardeners.
For more information visit *www.rhs.org.uk* or call 0845 130 4646.

RHS
LATIN
for
GARDENERS

Over 3,000 Plant Names
Explained and Explored

LORRAINE HARRISON

MITCHELL
BEAZLEY

CONTENTS

Preface		6
How to Use This Book		8
A Short History of Botanical Latin		9
Botanical Latin for Beginners		10
An Introduction to the A–Z Listings		13

Jasminum,
jasmine (p. 116)

THE A-Z LISTINGS OF LATIN PLANT NAMES

A	from *a-* to *azureus*	14
B	from *babylonicus* to *byzantinus*	37
C	from *cacaliifolius* to *cytisoides*	45
D	from *dactyliferus* to *dyerianum*	69
E	from *e-* to *eyriesii*	79
F	from *fabaceus* to *futilis*	85
G	from *gaditanus* to *gymnocarpus*	94
H	from *haastii* to *hystrix*	102
I	from *ibericus* to *ixocarpus*	109
J	from *jacobaeus* to *juvenilis*	115
K	from *kamtschaticus* to *kurdicus*	117
L	from *labiatus* to *lysimachioides*	118
M	from *macedonicus* to *myrtifolius*	129
N	from *nanellus* to *nymphoides*	141
O	from *obconicus* to *oxyphyllus*	146
P	from *pachy-* to *pyriformis*	150
Q	from *quadr-* to *quinquevulnerus*	173
R	from *racemiflorus* to *rutilans*	175
S	from *sabatius* to *szechuanicus*	181
T	from *tabularis* to *typhinus*	200
U	from *ulicinus* to *uvaria*	208
V	from *vacciniifolius* to *vulgatus*	212
W	from *wagnerii* to *wulfenii*	218
X	from *xanth-* to *xantholeucus*	220
Y	from *yakushimanus* to *yunnanensis*	220
Z	from *zabeliana* to *zonatus*	221

Tropaeolum majus,
nasturtium (p. 207)

Eryngium maritimum,
sea holly (p. 82)

Sir Joseph Banks,
1743–1820 (p. 40)

PLANT PROFILES

Acanthus	15
Achillea	16
Alyssum	23
Digitalis	76
Eryngium	82
Eucalyptus	84
Foeniculum	91
Geranium	95
Helianthus	103
Jasminum	116
Lycopersicon	128
Parthenocissus	153
Passiflora	154
Plumbago	163
Pulmonaria	171
Quercus	174
Sempervivum	188
Streptocarpus	195
Tropaeolum	207
Vaccinium	213

PLANT HUNTERS

Baron Alexander von Humboldt	26
Sir Joseph Banks	40
Meriwether Lewis and William Clark	54
Francis Masson and Carl Peter Thunberg	72
John Bartram and William Bartram	98
David Douglas	110
Carl Linnaeus	132
Jane Colden and Marianne North	158
Sir Joseph Hooker	182
André Michaux and François Michaux	210

PLANT THEMES

Where Plants Come From	32
Plants: Their Shape and Form	64
The Colour of Plants	86
The Qualities of Plants	120
Plants: Their Fragrance and Taste	144
Numbers and Plants	166
Plants and Animals	198
Glossary	222
Bibliography	223
Credits	224

Pelargonium,
pelargonium (p. 95)

PREFACE

When confronted with the complexities of botanical Latin many highly accomplished gardeners shrug, sigh and seek solace in Shakespeare. Just like Juliet they ask 'What's in a name? That which we call a rose / By any other name would smell as sweet.' Unfortunately, the subject cannot be dismissed quite that lightly. Doubtless the horticultural know-how of any self-respecting gardener would never let them mistake a dog rose (*Rosa canina*) for a rock rose (*Cistus ladanifer*), or a guelder rose (*Viburnum opulus*) for a Lenten rose (*Helleborus orientalis*). Only one of these very diverse plants bears any resemblance to Juliet's sweet-smelling rose, that which we commonly call the dog rose. The clue, of course, is in the Latin name which helpfully tells us it belongs to the genus *Rosa*, of the *Rosaceae* family, while *canina* pertains to dogs.

It is easy to understand why many of us fall for the poetry and charm of common names. Who can resist the romance of flowers with names like love-lies-bleeding, forget-me-not or love-in-a mist? (Although these pretty names can lead the gardener up the proverbial garden path as, for instance, the latter beauty also answers to the name devil-in-the-bush.) And after all, these florid appellations are so much easier to remember, and certainly to pronounce, than *Amaranthus caudatus*, *Myosotis sylvatica* and *Nigella damascena*. However poetic-sounding common names may be, they often tell us nothing about about the origin of a plant, or important things like its form, colour and size. When selecting plants it is so useful to know that those with *repens* in their name are low-growing or creeping, unlike those called *columnaris* which shoot up

tall. Think of the frustration avoided when planting a sunny bed if only one knew beforehand that *noctiflorus* flowers only open at night, and yet what helpful information to have if planning a garden primarily to be used in the evening.

The 18th-century botanist Carl Linnaeus introduced his simplified system of naming plants at a time when it was vital that physicians and herbalists could accurately identify and name plants, as these were the main source of medicine. Latin was the universal language of scholars and scientists and it is Linnaeus's binomial, or two-word, system that still forms the basis of the botanical Latin used by gardeners today. Thanks to Linnaeus and his colleagues a modern-day gardener in San Francisco can email a horticulturist in Hong Kong about *Chenopodium bonus-henricus* and know they are both discussing exactly the same plant. Could the same be said if their discussion relied on using one or more of its many common names that include goosefoot, shoemaker's heels or spearwort?

Far from being an esoteric or archaic language, when used appropriately botanical Latin can become a practical tool for creating a beautiful, productive and thriving garden that is quite as useful as a sharp pair of secateurs or a well-made trowel. Aided by this book the gardener can now answer the question 'What's in a name?' and they and their garden will benefit from understanding the wealth of information that has hitherto lain hidden within the mysterious world of Latin names.

Camellia × *williamsii* 'Citation',
(p. 219)

Lathyrus odoratus,
sweet pea (p. 145)

How to Use This Book

ALPHABETICAL LISTINGS

Latin terms appear alphabetically throughout for easy reference. For a more detailed explanation see the Introduction to the A–Z Listings on page 13.

The masculine form of the word always appears first, followed by the feminine (this is sometimes unchanged from the masculine) and then the neuter. In some cases, the epithet is the same for all three genders.

A guide to pronunciation is provided and capital letters indicate where the emphasis should fall.

abbreviatus *ab-bree-vee-AH-tus*
abbreviata, abbreviatum
Shortened; abbreviated, as in *Buddleja abbreviata*

When appropriate an example is given of a plant name which features the Latin term.

PLANT PROFILE PAGES

Throughout the book, feature pages focus on a particular plant, highlighting interesting and often entertaining Latin names, along with their derivations and sometimes surprising associations.

ACANTHUS

LATIN IN ACTION

Illustrated feature boxes reveal some of the hidden knowledge that can be derived from a plant's Latin name along with habitat information and useful cultivation tips.

PLANT HUNTERS

Read the stories of the men and women whose intrepid journeys resulted in the plant collections and introductions that have so shaped the way our gardens look today.

PLANT THEMES

In these pages Latin plant names are looked at thematically with such areas as the origin, colour, form and fragrance of plants covered. We also discover how plant names can often relate to numbers and to animals.

A Short History of Botanical Latin

The botanical Latin that scientists use today is quite distinct from the Latin of classical authors. It draws heavily on Greek and other languages, and uses many words that would have seemed barbarous to Roman writers on plants such as Pliny the Elder (23–79CE). Although its origins lie in the descriptive language used by those early botanists, it has developed into a specialised technical dialect, much simpler than classical Latin, but with a vocabulary that continues to expand to meet changing scientific requirements.

Until well into the 18th century, Latin was the language of international scholarship. It was therefore natural that botanists should use Latin in preference to vernacular plant names, which varied from language to language and region to region. From the 16th century onwards, pioneering voyages of discovery had resulted in a wealth of hitherto unknown plant material arriving in the studies of botanists throughout Europe. Technical developments in optical equipment meanwhile enabled far closer scrutiny of the structure of plants. Since Latin plant names were intended to encapsulate the differences between species, names often consisted of long strings of descriptive words, which were cumbersome to use and difficult to correlate. Then, in the mid-18th century, Carl Linnaeus (see p. 132) introduced his two-word, or binomial, system for naming plants and animals, by which a single epithet distinguished the species from all others in its genus.

This system transformed plant taxonomy. During the next century, it became clear that an internationally agreed set of rules for nomenclature was required. Through the international botanical congresses of the 19th and 20th centuries, this eventually led to the *International Code of Botanical Nomenclature* (ICBN), published in 1952, and revised several times since. This sets forth the principles by which plant names are formed and established; all major botanical journals and institutions abide by its rules and recommendations.

In view of which, it may seem strange that plant names are subject to so many changes. Gardeners may feel particularly aggrieved to have to learn a new name, when the old one seemed to do the job perfectly well. Unfortunately, botanists do not always agree on the relationship of one plant to another, and where conflicting classifications arise, changes of name may be the consequence. For instance, once evidence emerged that the genera *Cimicifuga* and *Actaea* were more closely related than previously thought, the plants that gardeners were accustomed to call *Cimicifuga* had to be re-learnt as *Actaea*.

The name *Actaea* was chosen over *Cimicifuga* because of the principle of priority, which is laid out in the ICBN. This states that where two entities are judged to be the same, the first name published must be used. Other consequences of taxonomic change can be equally confusing. For example, when species in *Montbretia* were reclassified as *Crocosmia*, the plant previously named *Montbretia × crocosmiiflora* (the montbretia with flowers like a crocosmia) became *Crocosmia × crocosmiiflora* (the crocosmia with flowers like a crocosmia).

Reclassification of plants has proceeded with even greater intensity in the era of DNA analysis, leading to significant numbers of name changes. The good news for gardeners is that the certainty with which DNA analysis allows relationships to be stated should eventually result in a far more robust taxonomy, where plant names will become more stable than ever before.

BOTANICAL LATIN
FOR BEGINNERS

When writing Latin names, it is important to order the various elements in the correct sequence and observe typographical conventions.

Family

(for example, *Sapindaceae*)
This appears as upper- and lower-case, and the *International Code of Botanical Nomenclature* also recommends italics. Family names are easily recognised, as they end in *-aceae*.

Genus

(for example, *Acer*)
This appears in italics with an upper-case initial letter. It is a noun and has a gender: masculine, feminine or neuter. The plural term is genera. When listing several species of the same genus together, the genus name is often abbreviated, for example: *Acer amoenum*, *A. barbinerve* and *A. calcaratum*.

Plumbago indica (syn. *P. rosea*), scarlet leadwort (p. 163)

Species

(for example, *Acer palmatum*)
The species is a specific unit within the genus and the term is often referred to as the specific epithet. This appears in lower-case italics. It is mostly an adjective, but can sometimes be a noun (for example, *Agave potatorum*, where the epithet means 'of drinkers'). Adjectives always agree in gender with the noun they follow, but nouns used as specific epithets are invariable. It is the combination of the generic and specific epithet that gives us the species name in the binomial, or two-word, system.

Subspecies

(for example, *Acer negundo* subsp. *mexicanum*)
This appears as lower-case italics and is preceded by the abbreviated form subsp. (or occasionally ssp.), which appears as lower-case roman type. It is a distinct variant of the main species.

Varietas

(for example, *Acer palmatum* var. *coreanum*)
This appears as lower-case italics and is preceded by the abbreviated form var., which appears as lower-case roman type. Also known as the variety, it is used to recognise slight variations in botanical structure.

Forma

(for example, *Acer mono* f. *ambiguum*)
This appears as lower-case italics and is preceded by the abbreviated form f., which appears in lower-case roman type. Also known as the form, it distinguishes minor variations such as the colour of the flower.

Cultivar

(for example, *Acer forrestii* 'Alice')
This appears as upper- and lower-case roman type
with single quotation marks. Also known as a
named variety, it is applied to artificially maintained
plants. Modern cultivar names (i.e., those after 1959)
should not include Latin or Latinised words.

Hybrid

(for example, *Hamamelis × intermedia*)
This appears as upper- and lower-case italics and is
preceded by a × that is not italicised (note that this
is a multiplication sign, not the letter x). This may
be applied to plants that are the product of a cross
between species of the same genus. If a hybrid
results from the crossing of species from different
genera then the hybrid generic name is preceded
by a multiplication sign. However, if a hybrid results
from the grafting of species, this is indicated by an
addition sign rather than a multiplication sign.

Synonyms

(for example, *Plumbago indica*, syn. *P. rosea*)
Within a classification a plant has only one correct
name, but it may have several incorrect ones.
These are known as synonyms (abbreviated as syn.)
and may have arisen due to two or more botanists
giving the same plant different names or from a
plant being classified in different ways.

Common names

Where common names are used, they appear as
lower-case roman type except where they derive from
a proper name, such as that of a person or a place.
(For example, common soapwort, but London pride.)
Note that Latinised versions of proper names do not
have capital letters, for example, *forrestii* or *freemanii*.
Many Latin genus names are in ordinary use as
common names, for example fuchsia; in this context,
they appear in lower-case roman type, and can also
be used in the plural (fuchsias, rhododendrons).

Helianthemum cupreum,
rock rose (p. 67)

Gender

In Latin, adjectives must agree with the gender of
the noun which they qualify; therefore in botanical
names, the species must agree with the genus.
An exception to this rule is made if the species name
is a noun (for example, *forrestii*, 'of Forrest'); in this
instance, there is no genus and species agreement.
To help familiarise the reader with the different gender
forms, where a specific epithet appears, the masculine,
feminine and neuter versions are usually listed –
for example, *grandiflorus* (*grandiflora*, *grandiflorum*).

This is a simplified outline of the binomial system,
but do beware as unfortunately exceptions to these
rules, along with further complexities of structure,
abound. As this is a book aimed at gardeners, not
botanists, and is not a Latin primer, only the broad
principles are dealt with here. The primary purpose
of this work is to encourage the blossoming of
better gardeners, not of Latin scholars. Helped by
an informed understanding of botanical nomenclature, it is hoped that gardeners will be able to make
better gardens filled with better plants; plants that
sit well in their site, thrive in the conditions provided
for them, and have the form, habit and colour that
make for the most aesthetically pleasing association
with their neighbours.

An Introduction to the A-Z Listings

Here follow more than 3,000 Latin botanical terms arranged alphabetically. The term appears first, then a guide to pronunciation. If applicable, the feminine and neuter versions of the name are then given, followed by the definition. An example of a plant name that features the term is also supplied. For example:

> abbreviatus *ab-bree-vee-AH-tus*
> abbreviata, abbreviatum
> Shortened; abbreviated, as in *Buddleja abbreviata*

For the sake of clarity and consistency, where the feminine term remains unchanged from the masculine it is repeated, as in:

> baicalensis *by-kol-EN-sis*
> baicalensis, baicalense
> From Lake Baikal, eastern Siberia, as in *Anemone baicalensis*

Where variations in spelling occur, they are grouped together, as in:

> cashmerianus *kash-meer-ee-AH-nus*
> cashmeriana, cashmerianum
> cashmirianus *kash-meer-ee-AH-nus*
> cashmiriana, cashmirianum
> cashmiriensis *kash-meer-ee-EN-sis*
> cashmiriensis, cashmiriense
> From or of Kashmir, as in *Cupressus cashmeriana*

Prosthechea vitellina. The species epithet (*vitellinus, vitellina, vitellinum*, yolk-yellow) describes the colour of the lip. (p. 217)

Pronunciation of botanical Latin can vary from country to country, and indeed from region to region, and the examples provided here are offered as a guide rather than a definitive directive. Capital letters indicate where the stress should fall. Where gender variations occur, only the masculine version is given.

Most gardeners will have encountered plants that seem to have not only numerous common names but also variations in their Latin appellations. Some epithets which were once in wide usage have now become obsolete, perhaps due to re-classification of the plants to which they were once applied. However, as these names may still sometimes be found in old horticultural works, as synonyms in modern texts or when browsing the varied sources of the internet they have been included for completeness.

It can be difficult to know which names are considered the most up-to-date, especially when different sources offer conflicting advice. Which classification is followed is largely a matter of personal choice but the *RHS Plant Finder* provides a general guide to current nomenclatural opinion, if one is sought.

Quercus suber,
cork oak (p. 174)

A

a-
Used in compound words to denote without or contrary to

abbreviatus *ab-bree-vee-AH-tus*
abbreviata, abbreviatum
Shortened; abbreviated, as in *Buddleja abbreviata*

abies *A-bees*
abietinus *ay-bee-TEE-nus*
abietina, abietinum
Like fir tree (*Abies*), as in *Picea abies*

abortivus *a-bor-TEE-vus*
abortiva, abortivum
Incomplete; with parts missing, as in *Oncidium abortivum*

abrotanifolius *ab-ro-tan-ih-FOH-lee-us*
abrotanifolia, abrotanifolium
With leaves like southernwood (*Artemisia abrotanum*),
as in *Euryops abrotanifolius*

abyssinicus *a-biss-IN-ih-kus*
abyssinica, abyssinicum
Connected with Abyssinia (Ethiopia), as in *Aponogeton abyssinicus*

Gentiana acaulis,
trumpet gentian

acanth-
Used in compound words to denote spiny, spiky or
thorny properties

acanthifolius *a-kanth-ih-FOH-lee-us*
acanthifolia, acanthifolium
With leaves like *Acanthus*, as in *Carlina acanthifolia*

acaulis *a-KAW-lis*
acaulis, acaule
Short-stemmed: without a stem, as in *Gentiana acaulis*

-aceae
Denoting the rank of family

acer *AY-sa*
acris, acre
With a sharp or pungent taste, as in *Sedum acre*

acerifolius *a-ser-ih-FOH-lee-us*
acerifolia, acerifolium
With leaves like maple (*Acer*), as in *Quercus acerifolia*

acerosus *a-seh-ROH-sus*
acerosa, acerosum
Like a needle, as in *Melaleuca acerosa*

acetosella *a-kee-TOE-sell-uh*
With slightly sour leaves, as in *Oxalis acetosella*

achilleifolius *ah-key-lee-FOH-lee-us*
achilleifolia, achilleifolium
With leaves like common yarrow (*Achillea millefolium*),
as in *Tanacetum achilleifolium*

acicularis *ass-ik-yew-LAH-ris*
acicularis, aciculare
Shaped like a needle, as in *Rosa acicularis*

acinaceus *a-sin-AY-see-us*
acinacea, acinaceum
In the shape of a curved sword or scimitar, as in *Acacia acinacea*

acmopetala *ak-mo-PET-uh-la*
With pointed petals, as in *Fritillaria acmopetala*

aconitifolius *a-kon-eye-tee-FOH-lee-us*
aconitifolia, aconitifolium
With leaves like aconite (*Aconitum*), as in
Ranunculus aconitifolius

ACANTHUS

The lush foliage and tall architectural flower spikes of the *Acanthus* plant strike a dramatic note in any garden. Belonging to the family *Acanthaceae*, the name for this genus of herbaceous perennials derives from *akantha*, the Greek for thorn. Where you see that *acanth-* forms part of the name of a plant then watch out, as it indicates that in some part it is spiny, spiky or thorny. For instance, the term *acanthocomus* (*acanthocoma, acanthocomum*) tells us the plant has spiny hairs on its leaves, while *acanthifolius* (*acanthifolia, acanthifolium*) means the leaves resemble those of the *Acanthus* plant. In Greek mythology, the nymph Acantha was much desired by the god Apollo. In an attempt to fight off his unwanted amorous advances, Acantha scratched Apollo's face; thus rejected, his revenge was to turn her into a spiky plant.

Acanthus, bear's breeches

Thwarted love apart, the spikiness associated with the *Acanthus* can also refer to the plant's flowers, which are formed from mauve and white overlapping bracts and tubular petals. These are borne on tall spikes that rise up gracefully from a mat of large leaves. This plant is often commonly called bear's breech, or bear's breeches. Among the most commonly grown is *Acanthus spinosus*, which has pointed, spiny leaves and produces an abundance of flowers, and can easily reach a stately 1.2 m (4 ft) – hence its common name spiny bear's breeches. *Acanthus montanus* is known as the mountain thistle (*spinosus, spinosa, spinosum* means spiny and *montanus, montana, montanum* means pertaining to mountains).

Acanthus plants thrive in dry, sunny spots in the garden, but be careful where you plant them, as they form a long taproot that makes it very difficult to remove them from unsuitable locations. They are generally hardy, but it is advisable to apply a generous mulch to cover the cut-down stems over the first couple of winters after planting.

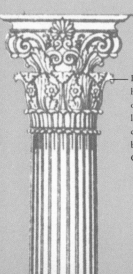

In classical architecture, highly stylised versions of the acanthus's curved leaf form part of the carved decoration on both Corinthian and Composite capitals.

According to the Roman poet Virgil, Helen of Troy wore a veil heavily embroidered with *Acanthus* leaves. More recently, the leaves appeared as a recurring motif in the work of the 19th-century artist and designer William Morris.

ACHILLEA

Named after the hero of Greek mythology, the Greek warrior Achilles, few plants can boast such an illustrious namesake as *Achillea*. Achilles's mother Thetis dipped the young Achilles in the magical waters of the River Styx, thus ensuring him protection from future attack. Held by his mother's hand, the only spot on his body that escaped immersion was his heel; needless to say, this is the point where the arrow from the bow aimed by his enemy Paris hit, killing the adult Achilles.

Before this final battle, Achilles had led many campaigns and was renowned for staunching the flow of blood from the wounds of his soldiers with a concoction prepared from yarrow (once known as *herba militaris*). The long-held belief that the plant had the power to heal wounds has resulted in a whole range of common names alluding to its medicinal properties, including bloodwort, nosebleed, staunchgrass and woundwort.

In more recent times, the white-flowered *Achillea ptarmica* has been used as snuff due to its sneeze-inducing qualities, and accounts for the common names sneezewort and old man's pepper (*ptarmica*, meaning to cause sneezing). Along with devil's nettle, *A. millefolium* is also known as thousand-leaf or thousand-seal, alluding to the great number of leaflets the plant produces, though probably not as many as 1,000. (*Millefolius*, *millefolia*, *millefolium* literally means with a thousand leaves, but usually simply signifies many leaves.)

Achillea millefolium,
yarrow

Achillea millefolium var. *rosea* – the name alludes to its rose-coloured flowers.

Achillea erba-rotta subsp. *moschata*, musk yarrow ; *erba-rotta* is a corruption of *herba rota*, from *rota*, a wheel.

The plentiful leaves of *Achillea* have a strong pungent smell and can cause sneezing in the unsuspecting gardener.

This group of hardy herbaceous perennials has once again become popular with gardeners as they team very well with ornamental grasses in mixed plantings, contributing stature, colour and plentiful, long-lived flowers to a border. Their flat flower heads composed of many tiny individual flowers rise from a mass of ferny leaves. They do best in well-drained soils in a sunny position. The taller cultivars should be cut hard to the ground in late autumn.

acraeus *ak-ra-EE-us*
acraea, acraeum
Dwelling on high ground, as in *Euryops acraeus*

actinophyllus *ak-ten-oh-FIL-us*
actinophylla, actinophyllum
With radiating leaves, as in *Schefflera actinophylla*

acu-
Used in compound words to denote sharply pointed

aculeatus *a-kew-lee-AH-tus*
aculeata, aculeatum
Prickly, as in *Polystichum aculeatum*

aculeolatus *a-kew-lee-oh-LAH-tus*
aculeolata, aculeolatum
With small prickles, as in *Arabis aculeolata*

acuminatifolius *a-kew-min-at-ih-FOH-lee-us*
acuminatifolia, acuminatifolium
With leaves that taper sharply to long narrow points, as in
Polygonatum acuminatifolium

acuminatus *ah-kew-min-AH-tus*
acuminata, acuminatum
Tapering to a long, narrow point, as in *Magnolia acuminata*

acutifolius *a-kew-ti-FOH-lee-us*
acutifolia, acutifolium
With leaves that taper quickly to sharp points, as in
Begonia acutifolia

acutilobus *a-KEW-ti-low-bus*
acutiloba, acutilobum
With sharply pointed lobes, as in *Hepatica acutiloba*

acutissimus *ak-yoo-TISS-ee-mum*
acutissima, acutissimum
With a very acute point, as in *Ligustrum acutissimum*

acutus *a-KEW-tus*
acuta, acutum
With a sharp but not tapering point, as in
Cynanchum acutum

ad-
Used in compound words to denote to

aden-
Used in compound words to denote that a part of the plant
has glands

adenophorus *ad-eh-NO-for-us*
adenophora, adenophorum
With glands, usually in reference to nectar, as in *Salvia adenophora*

adenophyllus *ad-en-oh-FIL-us*
adenophylla, adenophyllum
With sticky (gland-bearing) leaves, as in *Oxalis adenophylla*

adenopodus *a-den-OH-poh-dus*
adenopoda, adenopodum
With sticky pedicels (small stalks), as in *Begonia adenopoda*

adiantifolius *ad-ee-an-tee-FOH-lee-us*
adiantifolia, adiantifolium
With leaves like maidenhair fern (*Adiantum*), as in
Anemia adiantifolia

adlamii *ad-LAM-ee-eye*
Named after Richard Wills Adlam (1853–1903), a British collector
who supplied plants to London's Kew Gardens in the 1890s

admirabilis *ad-mir-AH-bil-is*
admirabilis, admirabile
Of note, as in *Drosera admirabilis*

Passiflora adiantifolia,
Norfolk Island passion flower

adnatus *ad-NAH-tus*
adnata, adnatum
Joined together, as in *Sambucus adnata*

adpressus *ad-PRESS-us*
adpressa, adpressum
Pressed close to; refers to the way hairs (for example) press against a
stem, as in *Cotoneaster adpressus*

adscendens *ad-SEN-denz*
Ascending; rising, as in *Aster adscendens*

adsurgens *ad-SER-jenz*
Rising upwards, as in *Phlox adsurgens*

aduncus *ad-UN-kus*
adunca, aduncum
Hooked, as in *Viola adunca*

aegyptiacus *eh-jip-tee-AH-kus*
aegyptiaca, aegyptiacum
aegypticus *eh-JIP-tih-kus*
aegyptica, aegypticum
aegyptius *eh-JIP-tee-us*
aegyptia, aegyptium
Connected with Egypt, as in *Achillea aegyptiaca*

aemulans *EM-yoo-lanz*
aemulus *EM-yoo-lus*
aemula, aemulum
Imitating; rivalling, as in *Scaevola aemula*

aequalis *ee-KWA-lis*
aequalis, aequale
Equal, as in *Phygelius aequalis*

aequinoctialis *eek-wee-nok-tee-AH-lis*
aequinoctialis, aequinoctiale
Connected with the equatorial regions, as in
Cydista aequinoctialis

aequitrilobus *eek-wee-try-LOH-bus*
aequitriloba, aequitrilobum
With three equal lobes, as in *Cymbalaria aequitriloba*

aerius *ER-re-us*
aeria, aerium
From high altitudes, as in *Crocus aerius*

aeruginosus *air-oo-jin-OH-sus*
aeruginosa, aeruginosum
The colour of rust, as in *Curcuma aeruginosa*

aesculifolius *es-kew-li-FOH-lee-us*
aesculifolia, aesculifolium
With leaves like horse chestnut (*Aesculus*), as in *Rodgersia aesculifolia*

aestivalis *ee-stiv-AH-lis*
aestivalis, aestivale
Relating to summer, as in *Vitis aestivalis*

aestivus *EE-stiv-us*
aestiva, aestivum
Developing or ripening in the summer months, as in
Leucojum aestivum

aethiopicus *ee-thee-OH-pih-kus*
aethiopica, aethiopicum
Connected with Africa, as in *Zantedeschia aethiopica*

aethusifolius *e-thu-si-FOH-lee-us*
aethusifolia, aethusifolium
With pungent leaves like *Aethusa*, as in *Aruncus aethusifolius*

Triticum aestivum,
wheat

aetnensis *eet-NEN-sis*
aetnensis, aetnense
From Mount Etna, Italy, as in *Genista aetnensis*

aetolicus *eet-OH-lih-kus*
aetolica, aetolicum
Connected with Aetolia, Greece, as in *Viola aetolica*

afer *a-fer*
afra, afrum
Specifically connected with North African coastal countries
such as Algeria and Tunisia, as in *Lycium afrum*

affinis *uh-FEE-nis*
affinis, affine
Related or similar to, as in *Dryopteris affinis*

afghanicus *af-GAN-ih-kus*
afghanica, afghanicum
afghanistanica *af-gan-is-STAN-ee-ka*
Connected with Afghanistan, as in *Corydalis afghanica*

aflatunensis *a-flat-u-NEN-sis*
aflatunensis, aflatunense
From Aflatun, Kyrgyzstan, as in *Allium aflatunense*

africanus *af-ri-KAHN-us*
africana, africanum
Connected with Africa, as in *Sparrmannia africana*

agastus *ag-AS-tus*
agasta, agastum
With great charm, as in *Rhododendron* × *agastum*

agavoides *ah-gav-OY-deez*
Resembling *Agave*, as in *Echeveria agavoides*

ageratifolius *ad-jur-rat-ih-FOH-lee-us*
ageratifolia, ageratifolium
With leaves like *Ageratum*, as in *Achillea ageratifolia*

ageratoides *ad-jur-rat-OY-deez*
Resembling *Ageratum*, as in *Aster ageratoides*

aggregatus *ag-gre-GAH-tus*
aggregata, aggregatum
Denotes aggregate flowers or fruits, such as raspberry or strawberry,
as in *Eucalyptus aggregata*

agnus-castus *AG-nus KAS-tus*
From *agnos*, Greek name for *Vitex agnus-castus*, and *castus*, chaste, as
in *Vitex agnus-castus*

Although *Saxifraga aizoides* has the common name
yellow saxifrage (or yellow mountain saxifrage), very
occasionally its flowers are dark red or orange. It is
found in North America and Europe and thrives in
cool, damp and rocky places. Insects love its bright
flowers, which are set amid fleshy leaves.

Saxifraga aizoides,
yellow saxifrage

agrarius *ag-RAH-ree-us*
agraria, agrarium
From fields and cultivated land, as in *Fumaria agraria*

agrestis *ag-RES-tis*
agrestis, agreste
Found growing in fields, as in *Fritillaria agrestis*

agrifolius *ag-rih-FOH-lee-us*
agrifolia, agrifolium
With leaves with a rough or scabby texture, as in *Quercus agrifolia*

agrippinum *ag-rip-EE-num*
Named after Agrippina, mother of the Emperor Nero, as in
Colchicum agrippinum

aitchisonii *EYE-chi-soh-nee-eye*
Named after Dr J.E.T. Aitchison (1836–98), a British doctor
and botanist who collected plant material in Asia, as in
Corydalis aitchisonii

aizoides *ay-ZOY-deez*
Like the genus *Aizoon*, as in *Saxifraga aizoides*

ajacis *a-JAY-sis*
A species name that honours the Greek hero Ajax,
as in *Consolida ajacis*

ajanensis *ah-yah-NEN-sis*
ajanensis, ajanense
From Ajan on the Siberian coast, as in *Dryas ajanensis*

alabamensis *al-uh-bam-EN-sis*
alabamensis, alabamense
alabamicus *al-a-BAM-ih-kus*
alabamica, alabamicum
From or of Alabama, USA, as in *Rhododendron alabamense*

alaternus *a-la-TER-nus*
The Roman name for *Rhamnus alaternus*

alatus *a-LAH-tus*
alata, alatum
Winged, as in *Euonymus alatus*

albanensis *al-ba-NEN-sis*
albanensis, albanense
From St Albans, Hertfordshire, England, as in
Coelogyne × *albanense*

alberti *al-BER-tee*
albertianus *al-ber-tee-AH-nus*
albertiana, albertianum
albertii *al-BER-tee-eye*
Named after various people called Albert, such as Albert von Regel
(1845–1908), plant collector, as in *Tulipa albertii*

albescens *al-BES-enz*
Becoming white, as in *Kniphofia albescens*

albicans *AL-bih-kanz*
Off-white, as in *Hebe albicans*

albicaulis *al-bih-KAW-lis*
albicaulis, albicaule
With white stems, as in *Lupinus albicaulis*

albidus *AL-bi-dus*
albida, albidum
White, as in *Trillium albidum*

albiflorus *al-BIH-flor-us*
albiflora, albiflorum
With white flowers, as in *Buddleja albiflora*

The genus *Iris* (family *Iridaceae*) consists of a large
range of species, varieties and hybrids. These are available in such a wide and varied range of colours that it
is sometimes easy for the gardener to overlook the less
vivid and more subtle-coloured blooms. One example
is *Iris albicans*, cultivated since ancient times. Its
common name, cemetery iris, refers to the Muslim
tradition of planting it by gravesides. Originally
native to Yemen and Saudi Arabia, this plant dislikes
cold regions but will thrive and spread quickly in
warm climates in well-drained sites. As the seed is
sterile, propagation is by rhizome division. Its height
range is 40–60 cm (15–24 in). It flowers early and has
fragrant blooms that, as its Latin name suggests, have
a delicate off-white hue.

Iris albicans,
cemetery iris

albifrons *AL-by-fronz*
With white fronds, as in *Cyathea albifrons*

albomaculatus *al-boh-mak-yoo-LAH-tus*
albomaculata, albomaculatum
With white spots, as in *Asarum albomaculatum*

albomarginatus *AL-bow-mar-gin-AH-tus*
albomarginata, albomarginatum
With white margins, as in *Agave albomarginata*

albopictus *al-boh-PIK-tus*
albopicta, albopictum
With white hairs, as in *Begonia albopicta*

albosinensis *al-bo-sy-NEN-sis*
albosinensis, albosinense
Meaning white and from China, as in *Betula albosinensis*

albovariegatus *al-bo-var-ee-GAH-tus*
albovariegata, albovariegatum
Variegated with white, as in *Holcus mollis* 'Albovariegatus'

albulus *ALB-yoo-lus*
albula, albulum
Whitish in colour, as in *Carex albula*

albus *AL-bus*
alba, album
White, as in *Veratrum album*

alcicornis *al-kee-KOR-nis*
alcicornis, alcicorne
Palmate leaves that resemble the horns of the European elk (North American moose), as in *Platycerium alcicorne*

aleppensis *a-le-PEN-sis*
aleppensis aleppense
aleppicus *a-LEP-ih-kus*
aleppica, aleppicum
From Aleppo, Syria, as in *Adonis aleppica*

aleuticus *a-LEW-tih-kus*
aleutica, aleuticum
Connected with the Aleutian Islands, Alaska, as in *Adiantum aleuticum*

alexandrae *al-ex-AN-dry*
Named after Queen Alexandra (1844–1925), wife of Edward VII of England, as in *Archontophoenix alexandrae*

alexandrinus *al-ex-an-DREE-nus*
alexandrina, alexandrinum
Connected with Alexandria, Egypt, as in *Senna alexandrina*

algeriensis *al-jir-ee-EN-sis*
algeriensis, algeriense
From Algeria, as in *Ornithogalum algeriense*

algidus *AL-gee-dus*
algida, algidum
Cold; of high mountain regions, as in *Olearia algida*

alienus *a-LY-en-us*
aliena, alienum
A plant of foreign origin, as in *Heterolepis aliena*

alkekengi *al-KEK-en-jee*
From the Arabic for bladder cherry, as in *Physalis alkekengi*

alleghaniensis *al-leh-gay-nee-EN-sis*
alleghaniensis, alleghaniense
From the Allegheny Mountains, USA, as in *Betula alleghaniensis*

alliaceus *al-lee-AY-see-us*
alliacea, alliacum
Like *Allium* (onion or garlic), as in *Tulbaghia alliacea*

alliariifolius *al-ee-ar-ee-FOH-lee-us*
alliariifolia, alliariifolium
With leaves like *Alliaria*, as in *Valeriana alliariifolia*

allionii *al-ee-OH-nee-eye*
Named after Carlo Allioni (1728–1804), Italian botanist, as in *Primula allionii*

alnifolius *al-nee-FOH-lee-us*
alnifolia, alnifolium
With leaves like alder (*Alnus*), as in *Sorbus alnifolia*

aloides *al-OY-deez*
Resembling *Aloe*, as in *Lachenalia aloides*

aloifolius *al-oh-ih-FOH-lee-us*
aloifolia, aloifolium
With leaves like *Aloe*, as in *Yucca aloifolia*

alopecuroides *al-oh-pek-yur-OY-deez*
Like the genus *Alopecurus* (foxtail), as in *Pennisetum alopecuroides*

alpestris *al-PES-tris*
alpestris, alpestre
Of lower, usually wooded, mountain habitats, as in
Narcissus alpestris

alpicola *al-PIH-koh-luh*
Of high mountain habitats, as in *Primula alpicola*

alpigenus *AL-pi-GEE-nus*
alpigena, alpigenum
Of a mountainous region, as in *Saxifraga alpigena*

alpinus *al-PEE-nus*
alpina, alpinum
Of high, often rocky regions; from the Alps region of Europe, as in
Pulsatilla alpina

altaclerensis *al-ta-cler-EN-sis*
altaclerensis, altaclerense
From Highclere Castle, Hampshire, England, as in *Ilex × altaclerensis*

altaicus *al-TAY-ih-kus*
altaica, altaicum
Connected with the Altai Mountains, Central Asia, as in
Tulipa altaica

alternans *al-TER-nans*
Alternating, as in *Chamaedorea alternans*

alternifolius *al-tern-ee-FOH-lee-us*
alternifolia, alternifolium
With leaves that grow from alternating points of a stem, rather than
opposite each other, as in *Buddleja alternifolia*

althaeoides *al-thay-OY-deez*
Resembling hollyhock (formerly *Althaea*), as in *Convolvulus althaeoides*

altissimus *al-TISS-ih-mus*
altissima, altissimum
Very tall; the tallest, as in *Ailanthus altissima*

altus *AHL-tus*
alta, altum
Tall, as in *Sempervivum altum*

amabilis *am-AH-bih-lis*
amabilis, amabile
Lovely, as in *Cynoglossum amabile*

amanus *a-MAH-nus*
amana, amanum
Of the Amanus Mountains, Turkey, as in *Origanum amanum*

amaranthoides *am-ar-anth-OY-deez*
Resembling amaranth (*Amaranthus*), as in *Calomeria amaranthoides*

amarellus *a-mar-ELL-us*
amarella, amarellum
amarus *a-MAH-rus*
amara, amarum
Bitter, as in *Ribes amarum*

amaricaulis *am-ar-ee-KAW-lis*
amaricaulis, amaricaule
With a bitter-tasting stem, as in *Hyophorbe amaricaulis*

amazonicus *am-uh-ZOH-nih-kus*
amazonica, amazonicum
Connected with the Amazon River, South America, as in
Victoria amazonica

ambi-
Used in compound words to denote around

ambiguus *am-big-YOO-us*
ambigua, ambiguum
Uncertain or doubtful, as in *Digitalis ambigua*

Pulsatilla alpina

ALYSSUM

Aurinia saxatilis,
gold dust

Where 'wort' appears as part of the common name of a plant, it denotes that the plant was believed to have curative and medicinal properties. In previous times, *Alyssum*, or madwort, was thought to protect against madness and the dangerous bites of mad dogs! The Latin name derives from the Greek *a*, meaning 'not' or 'against', and *lyssa*, meaning 'madness'. *Alyssoides* means resembling *Alyssum*, for instance *Alyssoides utriculata*, the bladderpod (*utriculatus, utriculata, utriculatum*, like a bladder). In the language of flowers, *Alyssum* means worth beyond beauty.

Aurinia saxatilis originates from the cooler regions of eastern Europe and Russia. It is sometimes known by the common name gold dust.

This group of plants includes hardy annuals, herbaceous perennials and small evergreen shrubs. They require a sunny position with well-drained soil. If happy, they produce a mass of flowers. To keep them compact and neat, the foliage should be cut back hard after flowering; otherwise they become very straggly and unkempt-looking. The name of the yellow-flowered *Alyssum argenteum* refers to its attractive grey-green foliage (*argenteus, argentea, argenteum* meaning silver in colour). The dazzling pure white of the flowers of many species is reflected in several cultivar names such as 'Avalanche', 'Carpet of Snow' and 'Snow Crystals'.

Alyssum is an excellent plant for growing in rock gardens or in the crevices of dry walls. The clue to this can be seen in the synonym name for *Aurinia saxatilis*, which is *Alyssum saxatile*. *Saxatilis* (*saxatilis, saxatile*) tells us that a plant is of or connected with rocky places, whereas *saxicola* means growing in rocky habitats and *saxosus* (*saxosa, saxosum*) means very rocky.

Lobularia maritima was formerly identified as *Alyssum* and belongs to the same family, *Brassicaceae* – hence its common name, sweet alyssum. In America it also goes by the name of sweet Alice. This is a very low-growing, mound-forming plant that produces white, lilac or pink flowers. Its name comes from the Latin *lobulus*, a small pod, an allusion to its small lobe-like fruits. Beware that confusion can arise between *Lobularia* and *Alyssum* in plant listings.

Alyssum cuneifolium,
madwort

amblyanthus *am-blee-AN-thus*
amblyantha, amblyanthum
With a blunt flower, as in *Indigofera amblyantha*

ambrosioides *am-bro-zhee-OY-deez*
Resembling *Ambrosia*, as in *Cephalaria ambrosioides*

amelloides *am-el-OY-deez*
Resembling *Aster amellus* (from its Roman name), as in
Felicia amelloides

americanus *a-mer-ih-KAH-nus*
americana, americanum
Connected with North or South America, as in
Lysichiton americanus

amesianus *ame-see-AH-nus*
amesiana, amesianum
Named after Frederick Lothrop Ames (1835–93), horticulturist and
orchid grower, and Oakes Ames (1874–1950), supervisor of the
Arnold Arboretum and professor of botany, Harvard, USA, as in
Cirrhopetalum amesianum

amethystinus *am-eth-ih-STEE-nus*
amethystina, amethystinum
Violet, as in *Brimeura amethystina*

ammophilus *am-oh-FIL-us*
ammophila, ammophilum
Of sandy places, as in *Oenothera ammophila*

Ribes americanum,
American blackcurrant

amoenus *am-oh-EN-us*
amoena, amoenum
Pleasant; delightful, as in *Lilium amoenum*

amphibius *am-FIB-ee-us*
amphibia, amphibium
Growing both on land and in water, as in *Persicaria amphibia*

amplexicaulis *am-pleks-ih-KAW-lis*
amplexicaulis, amplexicaule
Clasping the stem, as in *Persicaria amplexicaulis*

amplexifolius *am-pleks-ih-FOH-lee-us*
amplexifolia, amplexifolium
Clasping the leaf, as in *Streptopus amplexifolius*

ampliatus *am-pli-AH-tus*
ampliata, ampliatum
Enlarged, as in *Oncidium ampliatum*

amplissimus *am-PLIS-ih-mus*
amplissima, amplissimum
Very large, as in *Chelonistele amplissima*

amplus *AMP-lus*
ampla, amplum
Large, as in *Epigeneium amplum*

amurensis *am-or-EN-sis*
amurensis, amurense
From the Amur River region, Asia, as in *Sorbus amurensis*

amygdaliformis *am-mig-dal-ih-FOR-mis*
amygdaliformis, amygdaliforme
Shaped like an almond, as in *Pyrus amygdaliformis*

amygdalinus *am-mig-duh-LEE-nus*
amygdalina, amygdalinum
Relating to almond, as in *Eucalyptus amygdalina*

amygdaloides *am-ig-duh-LOY-deez*
Resembling almond, as in *Euphorbia amygdaloides*

ananassa *a-NAN-ass-uh*
ananassae *a-NAN-ass-uh-ee*
With a fragrance like pineapple, as in *Fragaria* × *ananassa*

anatolicus *an-ah-TOH-lih-kus*
anatolica, anatolicum
Connected with Anatolia, Turkey, as in *Muscari anatolicum*

anceps *AN-seps*
Two-sided, ambiguous, as in *Laelia anceps*

ancyrensis *an-syr-EN-sis*
ancyrensis, ancyrense
From Ankara, Turkey, as in *Crocus ancyrensis*

andersonianus *an-der-soh-nee-AH-nus*
andersoniana, andersonianum
andersonii *an-der-SON-ee-eye*
Named after Dr Charles Lewis Anderson (1827–1910), American
botanist, as in *Arctostaphylos andersonii*

andicola *an-DIH-koh-luh*
andinus *an-DEE-nus*
andina, andinum
Connected with the Andes, South America, as in *Calceolaria andina*

andrachne *an-DRAK-nee*
andrachnoides *an-drak-NOY-deez*
From the Greek *andrachne* (strawberry tree), as in
Arbutus × andrachnoides

andraeanus *an-dree-AH-nus*
andraeana, andraeanum
andreanus *an-dree-AH-nus*
andreana, andreanum
Named after Édouard François André (1840–1911), French explorer,
as in *Gymnocalycium andreae*

androgynus *an-DROG-in-us*
androgyna, androgynum
With separate male and female flowers growing on the same spike,
as in *Semele androgyna*

androsaemifolius *an-dro-say-MEE-fol-ee-us*
androsaemifolia, androsaemifolium
With leaves like *Androsaemum*, as in *Apocynum androsaemifolium*
(note: *Androsaemum* is now listed under *Hypericum*).

androsaemus *an-dro-SAY-mus*
androsaema, androsaemum
With sap the colour of blood, as in *Hypericum androsaemum*

anglicus *AN-glih-kus*
anglica, anglicum
Connected with England, as in *Sedum anglicum*

The various common names of this lovely wild
flower include rosebay willow herb, French willow
and fireweed. The latter name alludes to the fact that
this is often one of the first plants to colonise land
following a volcanic eruption. Its Latin name refers
to its long, thin leaves.

Chamaenerion angustifolium,
rosebay willow herb

angularis *ang-yoo-LAH-ris*
angularis, angulare
angulatus *ang-yoo-LAH-tus*
angulata, angulatum
Angular in shape or form, as in *Jasminum angulare*

angulosus *an-gew-LOH-sus*
angulosa, angulosum
With several corners or angles, as in *Bupleurum angulosum*

angustatus *an-gus-TAH-tus*
angustata, angustatum
Narrow, as in *Arisaema angustatum*

angustifolius *an-gus-tee-FOH-lee-us*
angustifolia, angustifolium
With narrow leaves, as in *Pulmonaria angustifolia*

BARON ALEXANDER
VON HUMBOLDT

(1769–1859)

Baron Alexander von Humboldt is one of the key figures in the history of the natural sciences. A highly accomplished polymath, he formed pioneering ideas concerning what he called 'the unity of nature', based on his rigorous empirical observations of the nature of biology, geology and meteorology.

Born in Berlin, then capital of Prussia, the son of a Prussian Army officer, Friedrich Heinrich Alexander von Humboldt relinquished his father's ambitions for him to follow in his brother Wilhelm's footsteps and enter politics, in favour of the more exciting life of an explorer. As a young man, his

friendship with the German naturalist and travel writer Georg Forster (1754–94) was very influential and inspired much of his later work. Forster had been part of Captain James Cook's second voyage (1772–5) aboard HMS *Resolution* bound for South Africa and the Antarctic. He and Humboldt travelled widely in Europe together.

Determined to live the life of an intrepid explorer, Humboldt set about preparatory study of such diverse areas of learning as anatomy, astronomy, geology and foreign languages. He then gained permission from the Spanish rulers to travel to and explore their South American dominions. Humboldt was accompanied by the French botanist and explorer Aimé Bonpland (1773–1858); he later described the plants they collected in the region in his publication *Plantes equinoxiales*. Their extensive five-year-long trip took them to Cuba, Mexico, up the Magdalena River and across to the Cordilleras then on to Quito and Lima. They travelled the Orinoco River and went to the source of the Amazon, where they confirmed the theory that these two great rivers were indeed connected by waterways, notably the Casiquiare Canal. Much of their journey covered vast tracts of wild, uninhabited and largely unexplored terrain; consequently, their expedition was not without danger. After discovering and capturing live electric eels, both Humboldt and Bonpland suffered potentially fatal

Such illustrious names as Darwin, Goethe, Jefferson and Schiller can all be counted among Baron Alexander von Humboldt's many admirers.

electric shocks. Undeterred, they collected numerous geological, zoological and botanical specimens, including 12,000 examples of plants. Based on his studies of the geography of the continent, Humboldt arrived at the radical hypothesis that the coastlines of South America and Africa had once been joined as one land mass. Before leaving the Americas, Humboldt travelled north to Washington and spent time at the White House with the president, Thomas Jefferson.

On his eventual return to Europe, Humboldt settled in Paris and spent more than 20 years committing to paper the wide-ranging knowledge he had gained on the expedition. This amounted to a staggering 30 volumes and resulted in his work being celebrated as the second great scientific discovery of South America. Honours from several countries were conferred upon him. Then, in 1845, at the age of 76, he published in five volumes *Cosmos: Draft of a Physical Description of the World*, in which he aimed to unify all the then known areas of scientific knowledge. His methodology, which was based on exacting observation and accurate measurement, became known as 'Humboldtian Science'. With hindsight, Baron Alexander von Humboldt can be viewed as an early environmental conservationist who developed an interconnected and somewhat holistic view of the natural world and how it worked. As early as 1799, he was expressing alarm at the extensive felling of the tree *Cinchona* in South America, the bark of which contains the medicine quinine. He also studied the rich fertilising properties of guano (dried bird excrement) and was responsible for introducing its use as a manure in Europe.

Plants named in his honour are numerous and bear the species name *humboldtii*. They include the *Lilium humboldtii*, *Geranium humboldtii*, the cactus

Lilium humboldtii,
Humboldt lily

Native to the foothills of California, the distinctive and lovely Humboldt lily is now under threat from habitat destruction.

Mammillaria humboldtii and *Quercus humboldtii*, commonly known as the South American or Columbian oak. Today his memory is also celebrated at the Alexander von Humboldt Foundation in Bonn, Germany. In his own day, Humboldt was a great supporter and patron of the work of his fellow scientists; this organisation continues the pioneering spirit of his work, granting scientific research fellowships and awards. Around the world, its many international alumni are known as Humboldtians.

'[HUMBOLDT] WAS THE GREATEST SCIENTIFIC TRAVELLER WHO EVER LIVED'
Charles Darwin (1809–82)

angustus *an-GUS-tus*
angusta, angustum
Narrow, as in *Rhodiola angusta*

anisatus *an-ee-SAH-tus*
anisata, anisatum
anisodorus *an-ee-so-DOR-us*
anisodora, anisodorum
With the scent of anise (*Pimpinella anisum*), as in *Illicium anisatum*

anisophyllus *an-ee-so-FIL-us*
anisophylla, anisophyllum
With leaves of unequal size, as in *Strobilanthes anisophylla*

annamensis *an-a-MEN-sis*
annamensis, annamense
From Annam, Asia, as in *Viburnum annamensis*

annulatus *an-yoo-LAH-tus*
annulata, annulatum
With rings, as in *Begonia annulata*

annuus *AN-yoo-us*
annua, annuum
Annual, as in *Helianthus annuus*

anomalus *ah-NOM-uh-lus*
anomala, anomalum
Unlike the norm found in a genus, as in *Hydrangea anomala*

anosmus *an-OS-mus*
anosma, anosmum
Without scent, as in *Dendrobium anosmum*

antarcticus *ant-ARK-tih-kus*
antarctica, antarcticum
Connected with the Antarctic region, as in *Dicksonia antarctica*

anthemoides *an-them-OY-deez*
Resembling chamomile (Greek *anthemis*), as in
Rhodanthe anthemoides

anthyllis *an-THILL-is*
Like kidney vetch (Greek *anthyllis*), as in *Erinacea anthyllis*

antipodus *an-te-PO-dus*
antipoda, antipodum
antipodeum *an-te-PO-dee-um*
Connected with the Antipodes, as in *Gaultheria antipoda*

antiquorum *an-ti-KWOR-um*
Of the ancients, as in *Helleborus antiquorum*

antiquus *an-TIK-yoo-us*
antiqua, antiquum
Ancient; antique, as in *Asplenium antiquum*

antirrhiniflorus *an-tee-rin-IF-lor-us*
antirrhiniflora, antirrhiniflorum
With flowers like snapdragon (*Antirrhinum*), as in
Maurandella antirrhiniflora

antirrhinoides *an-tee-ry-NOY-deez*
Resembling snapdragon (*Antirrhinum*), as in
Keckiella antirrhinoides

apenninus *ap-en-NEE-nus*
apennina, apenninum
Connected with the Apennine Mountains, Italy, as in
Anemone apennina

apertus *AP-ert-us*
aperta, apertum
Open; exposed, as in *Nomocharis aperta*

apetalus *a-PET-uh-lus*
apetala, apetalum
Without petals, as in *Sagina apetala*

aphyllus *a-FIL-us*
aphylla, aphyllum
Having, or appearing to have, no leaves, as in
Asparagus aphyllus

apiculatus *uh-pik-yoo-LAH-tus*
apiculata, apiculatum
With a short, sharp point, as in *Luma apiculata*

apiferus *a-PIH-fer-us*
apifera, apiferum
Bearing bees, as in *Ophrys apifera*

apiifolius *ap-ee-FOH-lee-us*
apiifolia, apiifolium
With leaves like celery (*Apium*), as in *Clematis apiifolia*

apodus *a-POH-dus*
apoda, apodum
Without stalks, as in *Selaginella apoda*

appendiculatus *ap-pen-dik-yoo-LAH-tus*
appendiculata, appendiculatum
With appendages such as hairs, as in *Caltha appendiculata*

applanatus *ap-PLAN-a-tus*

applanata, applanatum

Flattened, as in *Sanguisorba applanata*

appressus *a-PRESS-us*

appressa, appressum

Pressed close against, as in *Carex appressa*

apricus *AP-rih-kus*

aprica, apricum

Open to, or liking, the sun, as in *Silene aprica*

apterus *AP-ter-us*

aptera, apterum

Without wings, as in *Odontoglossum apterum*

aquaticus *a-KWA-tih-kus*

aquatica, aquaticum

aquatalis *ak-wa-TIL-is*

aquatalis, aquatale

Growing in or near water, as in *Mentha aquatica*

aquifolius *a-kwee-FOH-lee-us*

aquifolia, aquifolium

Holly-leaved (from the Latin name for holly, *aquifolium*), as in *Mahonia aquifolium*

aquilegiifolius *ak-wil-egg-ee-FOH-lee-us*

aquilegiifolia, aquilegiifolium

With leaves like columbine (*Aquilegia*), as in *Thalictrum aquilegiifolium*

aquilinus *ak-will-LEE-nus*

aquilina, aquilinum

Like an eagle; aquiline, as in *Pteridium aquilinum*

arabicus *a-RAB-ih-kus*

arabica, arabicum

Connected with Arabia, as in *Coffea arabica*

arachnoides *a-rak-NOY-deez*

arachnoideus *a-rak-NOY-dee-us*

arachnoidea, arachnoideum

Like a spider's web, as in *Sempervivum arachnoideum*

aralioides *a-ray-lee-OY-deez*

Like *Aralia*, as in *Trochodendron aralioides*

araucana *air-ah-KAY-nuh*

Relating to the Arauco region in Chile, as in *Araucaria araucana*

arbor-tristis *ar-bor-TRIS-tis*

Latin for sad tree, as in *Nyctanthes arbor-tristis*

arborescens *ar-bo-RES-senz*

arboreus *ar-BOR-ee-us*

arborea, arboreum

A woody or tree-like plant, as in *Erica arborea*

arboricola *ar-bor-IH-koh-luh*

Living on trees, as in *Schefflera arboricola*

arbusculus *ar-BUS-kyoo-lus*

arbuscula, arbusculum

Like a small tree, as in *Daphne arbuscula*

arbutifolius *ar-bew-tih-FOH-lee-us*

arbutifolia, arbutifolium

With leaves like strawberry tree (*Arbutus*), as in *Aronia arbutifolia*

archangelica *ark-an-JEL-ih-kuh*

In reference to the Archangel Raphael, as in *Angelica archangelica*

Ilex aquifolium,
European holly

archeri *ARCH-er-eye*
Named after William Archer (1820–74), Australian botanist,
as in *Eucalyptus archeri*

arcticus *ARK-tih-kus*
arctica, arcticum
Connected with the Arctic region, as in *Lupinus arcticus*

arcuatus *ark-yoo-AH-tus*
arcuata, arcuatum
In the shape of a bow or arc, as in *Blechnum arcuatum*

arenarius *ar-en-AH-ree-us*
arenaria, arenarium
arenicola *ar-en-IH-koh-luh*
arenosus *ar-en-OH-sus*
arenosa, arenosum
Growing in sandy places, as in *Leymus arenarius*

arendsii *ar-END-see-eye*
Named after Georg Arends (1862–1952), German nurseryman,
as in *Astilbe × arendsii*

areolatus *ar-ee-oh-LAH-tus*
areolata, areolatum
Areolate, with surface divided into small areas, as in
Coprosma areolata

argentatus *ar-jen-TAH-tus*
argentata, argentatum
argenteus *ar-JEN-tee-us*
argentea, argenteum
Silver in colour, as in *Salvia argentea*

argent-
Used in compound words to denote silver

argenteomarginatus *ar-gent-eoh-mar-gin-AH-tus*
argenteomarginata, argenteomarginatum
With silver edges, as in *Begonia argenteomarginata*

argentinus *ar-jen-TEE-nus*
argentina, argentinum
Connected with Argentina, as in *Tillandsia argentina*

argophyllus *ar-go-FIL-us*
argophylla, argophyllum
With silver leaves, as in *Eriogonum argophyllum*

argutifolius *ar-gew-tih-FOH-lee-us*
argutifolia, argutifolium
With sharp-toothed leaves, as in *Helleborus argutifolius*

argutus *ar-GOO-tus*
arguta, argutum
With notched edges, as in *Rubus argutus*

argyraeus *ar-jy-RAY-us*
argyraea, argyraeum
argyreus *ar-JY-ree-us*
argyrea, argyreum
Silvery in colour, as in *Dierama argyreum*

argyro-
Used in compound words to denote silver

argyrocomus *ar-gy-roh-KOH-mus*
argyrocoma, argyrocomum
With silver hairs, as in *Astelia argyrocoma*

argyroneurus *ar-ji-roh-NOOR-us*
argyroneura, argyroneurum
With silver veins, as in *Fittonia argyroneura*

LATIN IN ACTION

The prefix *argent-* in a binomial name means silver in
colour, as in *argentatus* and *argenteus*. It is therefore
rather surprising to find that the scarlet, yellow or
cream-flowered cockscomb's correct name is *Celosia
argentea* var. *cristata*.

Celosia argentea var. *cristata*
cockscomb

argyrophyllus *ar-ger-o-FIL-us*
argyrophylla, argyrophyllum
With silver leaves, as in *Rhododendron argyrophyllum*

aria *AR-ee-a*
From the Greek *aria*, probably whitebeam, as in *Sorbus aria*

aridus *AR-id-us*
arida, aridum
Growing in dry places, as in *Mimulus aridus*

arietinus *ar-ee-eh-TEEN-us*
arietina, arietinum
In the shape of a ram's head; horned, as in *Cypripedium arietinum*

arifolius *air-ih-FOH-lee-us*
arifolia, arifolium
With leaves like *Arum*, as in *Persicaria arifolia*

aristatus *a-ris-TAH-tus*
aristata, aristatum
Bearded, as in *Aloe aristata*

aristolochioides *a-ris-toh-loh-kee-OY-deez*
Resembling *Aristolochia*, as in *Nepenthes aristolochioides*

arizonicus *ar-ih-ZON-ih-kus*
arizonica, arizonicum
Connected with Arizona, as in *Yucca arizonica*

armandii *ar-MOND-ee-eye*
Named after Armand David (1826–1900), French naturalist and missionary, as in *Pinus armandii*

armatus *arm-AH-tus*
armata, armatum
With thorns, spines or spikes, as in *Dryandra armata*

armeniacus *ar-men-ee-AH-kus*
armeniaca, armeniacum
Connected with Armenia, as in *Muscari armeniacum*

armenus *ar-MEE-nus*
armena, armenum
Connected with Armenia, as in *Fritillaria armena*

armillaris *arm-il-LAH-ris*
armillaris, armillare
Like a bracelet, as in *Melaleuca armillaris*

Sorbus aria,
whitebeam

arnoldianus *ar-nold-ee-AH-nus*
arnoldiana, arnoldianum
Connected with the Arnold Arboretum in Boston, Massachusetts, USA, as in *Abies × arnoldiana*

aromaticus *ar-oh-MAT-ih-kus*
aromatica, aromaticum
With a fragrant, aromatic scent, as in *Lycaste aromatica*

artemisioides *ar-tem-iss-ee-OY-deez*
Resembling *Artemisia*, as in *Senna artemisioides*

articulatus *ar-tik-oo-LAH-tus*
articulata, articulatum
With a jointed stem, as in *Senecio articulatus*

arundinaceus *a-run-din-uh-KEE-us*
arundinacea, arundinaceum
Like a reed, as in *Phalaris arundinacea*

arvensis *ar-VEN-sis*
arvensis, arvense
Growing in cultivated fields, as in *Rosa arvensis*

asarifolius *as-ah-rih-FOH-lee-us*
asarifolia, asarifolium
With leaves like wild ginger (*Asarum*), as in *Cardamine asarifolia*

ascendens *as-SEN-denz*
Rising upwards, as in *Calamintha ascendens*

asclepiadeus *ass-cle-pee-AD-ee-us*
asclepiadea, asclepiadeum
Like milkweed (*Asclepias*), as in *Gentiana asclepiadea*

Where Plants Come From

Many species names provide the gardener with helpful information concerning the origins of a plant. Often these names refer to the continent or country where the species was originally collected, thus telling us something about its native habitat. Armed with one or two clues about the natural location of a particular plant, the gardener can then begin to assess whether it might thrive or flounder once transplanted to his or her own plot, saving the heartache and expense caused by choosing inappropriate plants that were doomed from the start. However, the detail and specificity of the information these names provide can vary widely; it may be as generalised as a compass point or as exact as the location where a plant was bred.

Starting with the wider picture, *borealis* (*borealis*, *boreale*) means northern; *australis* (*australis*, *australe*) southern; *orientalis* (*orientalis*, *orientale*) indicates eastern, and *occidentalis* (*occidentalis*, *occidentale*) is western. Extra detail may be provided with terms such as *hyperborealis*, meaning the far north. Continents often form part of the Latin name of a plant, such as *africanus* (*africana*, *africanum* – Africa) and *europaeus* (*europaea*, *europaeum* – Europe), along with individual countries like Spain – *hispanicus* (*hispanica*, *hispanicum*) or Japan – *japonicus* (*japonica*, *japonicum*). Some countries have more than one epithet; for instance, Japan is also indicated by *nipponicus* (*nipponica*, *nipponicum*). In keeping with the rules of botanical Latin, these place names are never capitalised. The names of states, cities and towns frequently appear, such as *missouriensis* (*missouriensis*, *missouriense*), for the American state of Missouri, and *thebaicus* (*thebaica*, *thebaicum*), for the ancient Egyptian settlement of Thebes, now Luxor. One needs to be mindful of the vicissitudes of history, as such name changes can cause great confusion when trying to link a plant to a place. Broad geographical regions that cross country borders are also common and avoid some of the more obvious dangers associated with contested or altered boundaries; one such example is *aegeus* (*aegea*, *aegeum*), meaning connected with or of the shores of the Aegean Sea.

Perhaps most helpful to the gardener are species names that tell us about the specific growing conditions most favoured by a plant. Once you know that the natural habitat of plants with *ammophilus*

***Digitalis canariensis*,**
Canary Island foxglove

This plant is named after the group of islands in the Atlantic Ocean.

Verbena canadensis,
rose verbena

Although properly applied just to Canada, *canadensis*
often refers to north-eastern parts of the USA.

Baptisia australis,
silver sweet pea bush

Australis means southern, but can also simply indicate that a plant
has a more southerly distribution than other members of the genus.

(*ammophila, ammophilum*) in their name is sandy
places, or that *salinus* (*salina, salinum*) plants are
associated with salty regions, you will have a much
greater chance of finding an auspicious spot for them.
When these Latin terms become a little more
familiar, it is easy to extract useful guidelines to
such things as the most suitable climate, aspect or
altitude for a particular plant. The epithet *montanus*
(*montana, montanum*) means relating to mountains,
while *monticola* tells us that in the wild this plant
can be found growing on mountains, both terms
indicating a certain degree of hardiness. Plants native
to high and rocky regions, including the European
Alps, often have *alpinus* (*alpina, alpinum*) in their
nomenclature, such as the alpine aster, *Aster alpinus*,
whereas those of lower, usually wooded, mountain
habitats are known as *alpestris* (*alpestris, alpestre*).

Gardeners whose plots are bathed in continuous
daytime sunshine with freely draining soil would
do well to memorise the Latin word for a wood,
silva, then vow never to be tempted by plants with
similar-sounding names! *Sylvaticus* (*sylvatica,
sylvaticum*), *sylvestris* (*sylvestris, sylvestre*) and
sylvicola all indicate that a plant is of a woodland
or forest setting and will be happiest if planted
in similar conditions.

Perhaps less helpful are generalised geographical
indicators such as *accola*, which simply tells us that
a plant is from nearby without specifying where.
Occasionally, a term may have one or more meanings,
as in *peregrinus* (*peregrina, peregrinum*); this may
mean that a plant is exotic or foreign, or that it
is given to wandering, as in *Erigeron peregrinus*,
the wandering daisy.

Ranunculus asiaticus,
Persian buttercup

aselliformis *ass-el-ee-FOR-mis*
aselliformis, aselliforme
Shaped like a woodlouse, as in *Pelecyphora aselliformis*

asiaticus *a-see-AT-ih-kus*
asiatica, asiaticum
Connected with Asia, as in *Trachelospermum asiaticum*

asparagoides *as-par-a-GOY-deez*
Resembling or like asparagus, as in *Acacia asparagoides*

asper *AS-per*
aspera, asperum
asperatus *as-per-AH-tus*
asperata, asperatum
With a rough texture, as in *Hydrangea aspera*

asperifolius *as-per-ih-FOH-lee-us*
asperifolia, asperifolium
With rough leaves, as in *Cornus asperifolia*

asperrimus *as-PER-rih-mus*
asperrima, asperrimum
With a very rough texture, as in *Agave asperrima*

asphodeloides *ass-fo-del-oy-deez*
Like *Asphodelus*, as in *Geranium asphodeloides*

asplenifolius *ass-plee-ni-FOH-lee-us*
asplenifolia, asplenifolium
aspleniifolius *ass-plee-ni-eye-FOH-lee-us*
aspleniifolia, aspleniifolium
With fine, feathery, fern-like leaves, as in *Phyllocladus aspleniifolia*

assa-foetida *ass-uh-FET-uh-duh*
From the Persian *aza*, mastic, and Latin *foetidus*, stinking, as in *Ferula assa-foetida*

assimilis *as-SIM-il-is*
assimilis, assimile
Similar; alike, as in *Camellia assimilis*

assurgentiflorus *as-sur-jen-tih-FLOR-us*
assurgentiflora, assurgentiflorum
With flowers in ascending clusters, as in *Lavatera assurgentiflora*

assyriacus *ass-see-re-AH-kus*
assyriaca, assyriacum
Connected with Assyria, as in *Fritillaria assyriaca*

asteroides *ass-ter-OY-deez*
Resembling *Aster*, as in *Amellus asteroides*

astilboides *a-stil-BOY-deez*
Resembling *Astilbe*, as in *Astilbe astilboides*

asturiensis *ass-tur-ee-EN-sis*
asturiensis, asturiense
From the province of Asturias, Spain, as in *Narcissus asturiensis*

atkinsianus *at-kin-see-AH-nus*
atkinsiana, atkinsianum
atkinsii *at-KIN-see-eye*
Named after James Atkins (1802–84), English nurseryman, as in *Petunia × atkinsiana*

atlanticus *at-LAN-tih-kus*
atlantica, atlanticum
Connected with the Atlantic shoreline, or from the Atlas Mountains, as in *Cedrus atlantica*

atriplicifolius *at-ry-pliss-ih-FOH-lee-us*
atriplicifolia, atriplicifolium
With leaves like orache or salt-bush (*Atriplex*), as in *Perovskia atriplicifolia*

atro-
Used in compound words to denote dark

atrocarpus *at-ro-KAR-pus*
atrocarpa, atrocarpum
With black or very dark fruit, as in *Berberis atrocarpa*

atropurpureus *at-ro-pur-PURR-ee-us*
atropurpurea, atropurpureum
Dark purple, as in *Scabiosa atropurpurea*

atrorubens *at-roh-ROO-benz*
Dark red, as in *Helleborus atrorubens*

atrosanguineus *at-ro-san-GWIN-ee-us*
atrosanguinea, atrosanguineum
Dark blood red, as in *Rhodochiton atrosanguineus*

atroviolaceus *at-roh-vy-oh-LAH-see-us*
atroviolacea, atroviolaceum
Dark violet, as in *Dendrobium atroviolaceum*

atrovirens *at-ro-VY-renz*
Dark green, as in *Chamaedorea atrovirens*

attenuatus *at-ten-yoo-AH-tus*
attenuata, attenuatum
With a narrow point, as in *Haworthia attenuata*

atticus *AT-tih-kus*
attica, atticum
Connected with Attica, Greece, as in *Ornithogalum atticum*

aubrietioides *au-bre-teh-OY-deez*
aubrietiodes
Resembling *Aubrieta*, as in *Arabis aubrietiodes*

aucheri *aw-CHER-ee*
Named after Pierre Martin Rémi Aucher-Éloy (1792–1838),
French pharmacist and botanist, as in *Iris aucheri*

aucuparius *awk-yoo-PAH-ree-us*
aucuparia, aucuparium
Of bird-catching, as in *Sorbus aucuparia*

augustinii *aw-gus-TIN-ee-eye*
augustinei
Named after the Irish plantsman and botanist Dr Augustine Henry
(1857–1930), as in *Rhododendron augustinii*

augustissimus *aw-gus-TIS-sih-mus*
augustissima, augustissimum
augustus *aw-GUS-tus*
augusta, augustum
Majestic, noteworthy, as in *Abroma augusta*

aurantiacus *aw-ran-ti-AH-kus*
aurantiaca, aurantiacum
aurantius *aw-RAN-tee-us*
aurantia, aurantium
Orange, as in *Pilosella aurantiaca*

aurantiifolius *aw-ran-tee-FOH-lee-us*
aurantiifolia, aurantiifolium
With leaves like an orange tree (*Citrus aurantium*), as in
Citrus aurantiifolia

auratus *aw-RAH-tus*
aurata, auratum
With golden rays, as in *Lilium auratum*

aureo-
Used in compound words to denote golden

aureosulcatus *aw-ree-oh-sul-KAH-tus*
aureosulcata, aureosulcatum
With yellow furrows, as in *Phyllostachys aureosulcata*

aureus *AW-re-us*
aurea, aureum
Golden yellow, as in *Phyllostachys aurea*

LATIN IN ACTION

Atrorubens describes the deep purple-plum flowers of
this Lenten rose (also known as the Christmas rose),
which will add a touch of drama and glamour to any
late-winter garden border. Happiest in partial or light
shade and moist but well-drained soil, hellebores do
not like to be moved once established.

Helleborus atrorubens, Lenten rose

auricomus *aw-RIK-oh-mus*
auricoma, auricomum
With golden hair, as in *Ranunculus auricomus*

auriculatus *aw-rik-yoo-LAH-tus*
auriculata, auriculatum
auriculus *aw-RIK-yoo-lus*
auricula, auriculum
auritus *aw-RY-tus*
aurita, auritum
With ears or ear-shaped appendages, as in
Plumbago auriculata

australiensis *aw-stra-li-EN-sis*
australiensis, australiense
From Australia, as in *Idiospermum autraliense*

australis *aw-STRAH-lis*
australis, australe
Southern, as in *Cordyline australis*

austriacus *oss-tree-AH-kus*
austriaca, austriacum
Connected with Austria, as in *Doronicum austriacum*

austrinus *oss-TEE-nus*
austrina, austrinum
Southern, as in *Rhododendron austrinum*

autumnalis *aw-tum-NAH-lis*
autumnalis, autumnale
Relating to autumn, as in *Colchicum autumnale*

avellanus *av-el-AH-nus*
avellana, avellanum
Connected with Avella, Italy, as in *Corylus avellana*

avenaceus *a-vee-NAY-see-us*
avenacea, avenaceum
Like *Avena* (oats), as in *Agrostis avenacea*

avium *AY-ve-um*
Relating to birds, as in *Prunus avium*

axillaris *ax-ILL-ah-ris*
axillaris, axillare
Growing in the axil, as in *Petunia axillaris*

azedarach *az-ee-duh-rak*
From the Persian for noble tree, as in *Melia azedarach*

azoricus *a-ZOR-ih-kus*
azorica, azoricum
Connected with the Azores Islands, as in *Jasminum azoricum*

azureus *a-ZOOR-ee-us*
azurea, azureum
Azure; sky blue, as in *Muscari azureum*

LATIN IN ACTION

Belonging to the family *Boraginaceae*, species of the
genus *Anchusa* are among the very best blue-flowered
plants that the gardener can grow. Popular cultivars
of *Anchusa azurea* include the deep-blue 'Dropmore'
and gentian-blue 'Lodden Royalist'. Natives of Central
Asia and the Mediterranean regions, there are many
species that include hardy biennials and herbaceous
perennials. They thrive in full sun and fertile but
well-drained soil – add sand at the time of planting if
soil is heavy. *A. azurea* will establish readily from root
cuttings taken early in the year then planted out in
autumn; it grows to a height of over 1 m (3–4 ft). Root
division is the only way to ensure purity of colour, as
they may not come true from seed. Smaller species are
particularly good for rock gardens and can often be
seen growing in pans in alpine houses.

Anchusa azurea,
garden anchusa

B

babylonicus *bab-il-LON-ih-kus*
babylonica, babylonicum
Connected with Babylonia, Mesopotamia (Iraq), as in
Salix babylonica, which Linnaeus mistakenly believed to be from the
south-west Asia

baccans *BAK-kanz*
bacciferus *bak-IH-fer-us*
baccifera, bacciferum
With berries, as in *Erica baccans*

baccatus *BAK-ah-tus*
baccata, baccatum
With fleshy berries, as in *Malus baccata*

bacillaris *bak-ILL-ah-ris*
bacillaris, bacillare
Like a stick, as in *Cotoneaster bacillaris*

backhouseanus *bak-how-zee-AH-nus*
backhouseana, backhouseanum
backhousianus *bak-how-zee-AH-nus*
backhousiana, backhousianum
backhousei *bak-HOW-zee-eye*
Named after James Backhouse (1794–1869), British nurseryman,
as in *Correa backhouseana*

badius *bad-ee-AH-nus*
badia, badium
Chestnut brown, as in *Trifolium badium*

baicalensis *by-kol-EN-sis*
baicalensis, baicalense
From Lake Baikal, eastern Siberia, as in *Anemone baicalensis*

baileyi *BAY-lee-eye*
baileyanus *bay-lee-AH-nus*
baileyana, baileyanum
Named after one of the following:
Frederick Manson Bailey (1827–1915), Australian botanist;
Lt Colonel Frederick Marshman Bailey (1882–1967), Indian
Army soldier who collected plants on the Tibetan borders
from 1913; Major Vernon Bailey (1864–1942), American Army
soldier who collected cacti from 1900; Liberty Hyde Bailey
(1858–1954), author and professor of horticulture at Cornell
University, USA, as in *Rhododendron baileyi* (named after
Lt Colonel Frederick Marshman Bailey)

bakeri *BAY-ker-eye*
bakerianus *bay-ker-ee-AH-nus*
bakeriana, bakerianum
Usually honouring John Gilbert Baker (1834–1920) of Kew, as in
Aloe bakeri, but also George Percival Baker (1856–1951), British
plant collector

baldensis *bald-EN-sis*
baldensis, baldense
baldianus *bald-ee-AN-ee-us*
baldiana, baldianum
From or of Monte Baldo, Italy, as in *Gymnocalycium baldianum*

baldschuanicus *bald-SHWAN-ih-kus*
baldschuanica, baldschuanicum
Connected with Baljuan, Turkistan, as in *Fallopia baldschuanica*

balearicus *bal-AIR-ih-kus*
balearica, balearicum
Connected with the Balearic Islands, Spain, as in *Buxus balearica*

balsameus *bal-SAM-ee-us*
balsamea, balsameum
Like balsam, as in *Abies balsamea*

balsamiferus *bal-sam-IH-fer-us*
balsamifera, balsamiferum
Producing balsam, as in *Aeonium balsamiferum*

balticus *BUL-tih-kus*
baltica, balticum
Connected with the Baltic Sea region, as in *Cotoneaster balticus*

bambusoides *bam-BOO-soy-deez*
Resembling bamboo (*Bambusa*), as in *Phyllostachys bambusoides*

banaticus *ba-NAT-ih-kus*
banatica, banaticum
Connected with the Banat region of Central Europe,
as in *Crocus banaticus*

banksianus *banks-ee-AH-nus*
banksiana, banksianum
banksii *BANK-see-eye*
Named after Sir Joseph Banks (1743–1820), English botanist and
plant collector, as in *Cordyline banksii*; *banksiae* commemorates his
wife, Lady Dorothea Banks (1758–1828)

bannaticus *ban-AT-ih-kus*
bannatica, bannaticum
Connected with Banat, Central Europe, as in
Echinops bannaticus

barbarus *BAR-bar-rus*
barbara, barbarum
Foreign, as in *Lycium barbarum*

barbatulus *bar-BAT-yoo-lus*
barbatula, barbatulum
barbatus *bar-BAH-tus*
barbata, barbatum
Bearded; with long, weak hairs, as in *Hypericum barbatum*

barbigerus *bar-BEE-ger-us*
barbigera, barbigerum
With beards or barbs, as in *Bulbophyllum barbigerum*

barbinervis *bar-bih-NER-vis*
barbinervis, barbinerve
With bearded or barbed veins, as in *Clethra barbinervis*

barbinodis *bar-bin-OH-dis*
barbinodis, barbinode
With beards at the nodes or joints, as in *Bothriochloa barbinodis*

barbulatus *bar-bul-AH-tus*
barbulata, barbulatum
With a short or less significant beard, as in *Anemone barbulata*

barcinonensis *bar-sin-oh-NEN-sis*
barcinonensis, barcinonense
From Barcelona, Spain, as in *Galium × barcinonense*

baselloides *bar-sell-OY-deez*
Resembling *Basella*, as in *Boussangaultia baselloides*

basilaris *bas-il-LAH-ris*
basilaris, basilare
Relating to the base or bottom, as in *Opuntia basilaris*

basilicus *bass-IL-ih-kus*
basilica, basilicum
With princely or royal properties, as in *Ocimum basilicum*

baueri *baw-WARE-eye*
bauerianus *baw-ware-ee-AH-nus*
baueriana, bauerianum
Named after Ferdinand Bauer (1760–1826), Austrian botanical artist to Flinders' Australian expedition, as in *Eucalyptus baueriana*

baurii *BOUR-ee-eye*
Named after Dr Georg Herman Carl Ludwig Baur (1859–98), German plant collector, as in *Rhodohypoxis baurii*

beesianus *bee-zee-AH-nus*
beesiana, beesianum
Named after Bees Nursery, Chester, England, as in *Allium beesianum*

belladonna *bel-uh-DON-nuh*
Beautiful lady, as in *Amaryllis belladonna*

bellidifolius *bel-lid-ee-FOH-lee-us*
bellidifolia, bellidifolium
With leaves like a daisy (*Bellis*), as in *Ageratina bellidifolia*

bellidiformis *bel-id-EE-for-mis*
bellidiformis, bellidiforme
Like a daisy (*Bellis*), as in *Dorotheanthus bellidiformis*

bellidioides *bell-id-ee-OY-deez*
Resembling *Bellium*, as in *Silene bellidioides*

bellus *BELL-us*
bella, bellum
Beautiful; handsome, as in *Graptopetalum bellum*

benedictus *ben-uh-DICK-tus*
benedicta, benedictum
A blessed plant; spoken of favourably, as in *Centaurea benedicta*

benghalensis *ben-gal-EN-sis*
benghalensis, benghalense
Also *bengalensis*; from Bengal, India, as in *Ficus benghalensis*

bermudianus *ber-myoo-dee-AH-nus*
bermudiana, bermudianum
Connected with Bermuda, as in *Juniperus bermudiana*

berolinensis *ber-oh-lin-EN-sis*
berolinensis, berolinense
From Berlin, Germany, as in *Populus × berolinensis*

berthelotii *berth-eh-LOT-ee-eye*
Named after Sabin Berthelot (1794–1880), French naturalist,
as in *Lotus berthelotii*

betaceus *bet-uh-KEE-us*
betacea, betaceum
Like a beet (*Beta*), as in *Solanum betaceum*

betonicifolius *bet-on-ih-see-FOH-lee-us*
betonicifolia, betonicifolium
Like betony (*Stachys*), as in *Meconopsis betonicifolia*

betulifolius *bet-yoo-lee-FOH-lee-us*
betulifolia, betulifolium
With leaves like a birch (*Betula*), as in *Pyrus betulifolia*

betulinus *bet-yoo-LEE-nus*
betulina, betulinum
betuloides *bet-yoo-LOY-deez*
Resembling or like a birch (*Betula*), as in *Carpinus betulinus*

bicolor *BY-kul-ur*
With two colours, as in *Caladium bicolor*

Ficus benghalensis,
banyan

bicornis *BY-korn-is*
bicornis, bicorne
bicornutus *by-kor-NOO-tus*
bicornuta, bicornutum
With two horns or horn-like spurs, as in *Passiflora bicornis*

bidentatus *by-den-TAH-tus*
bidentata, bidentatum
With two teeth, as in *Allium bidentatum*

biennis *by-EN-is*
biennis, bienne
Biennial, as in *Oenothera biennis*

bifidus *BIF-id-us*
bifida, bifidum
Cleft in two parts, as in *Rhodophila bifida*

SIR JOSEPH BANKS

(1743–1820)

The legacy of Sir Joseph Banks is today evident in the widespread distribution around the world of a whole host of plants, many of which bear his name. The genus *Banksia*, which includes the Australian honeysuckle, is one of his most famous discoveries. Other plants named in his honour are recognised in the species names *banksii* or *banksianus*.

The young Banks displayed an early interest in the natural world; he spent many childhood hours botanising on his family estate of Revesby Abbey in Lincolnshire, England. After the death of his father, his mother moved to London, in the vicinity of the Chelsea Physic Garden, whose collection of exotic plants greatly inspired the young Banks. He read botany at Christ Church, Oxford, but did not complete his studies, opting instead for the far more exciting prospect of joining a voyage of discovery aboard HMS *Niger*. Travelling to Newfoundland and Labrador in 1766–7, Banks collected plants, rocks and animals, and introduced *Rhododendron canadense* into Britain. Interestingly, it was the volcanic lava that Banks collected in Iceland and later gave to the Chelsea Physic Garden that was to form the basis of the first rock garden created in Europe.

On his return to England, Banks was invited to join Captain James Cook's *Endeavour* expedition to the South Pacific, which set sail the following year. Already a Fellow of the Royal Society, Banks was ably

Banks had a long and distinguished career in botany, both in the field and as a president of the Royal Society. He was also a founder member of the Horticultural Society in 1804 (now known as the Royal Horticultural Society).

assisted by Daniel Solander (1733–82); they spent the three-year trip studying and recording the natural history of South America, Tahiti, New Zealand, Australia and Java. On this trip alone, he secured more than 3,500 dried plant specimens, 1,400 of which were new to science. Among the species he introduced into European cultivation were *Leptospermum scoparium* from New Zealand and *Eucalyptus gummifera* and *Dianella caerulea* from Australia. While exploring the east coast of Australia, Banks suggested that Cook should name a particularly species-rich site Botany Bay; *Banksia serrata* is thought to be the first plant grown in England from seed collected in that area. Banks even managed to collect around 400 coastal plants from the shores of New Zealand, despite coming under attack from aggressive and cannibalistic Maoris. Banks's final voyage of exploration was to Iceland in 1772; again, Solander accompanied him as assistant. Introductions from this voyage included *Koenigia islandica* and *Salix myrtilloides*.

Although most of the rest of his life was spent on his native shores, Sir Joseph Banks continued to add greatly to the botanical knowledge of the period and exercised a profound influence on the transportation of plants around the world. From the early 1770s, he acted as unofficial director of London's Kew Gardens; it was in this capacity that he was responsible for instigating several important plant-hunting expeditions, sending Kew botanists such as Francis Masson to several continents. Notable among these was Archibald Menzies' 1791 trip to the northwest coast of the USA and William Kerr's expedition to China. This latter expedition resulted in the introduction of *Magnolia denudata*. Such voyages continued well into the 19th century and included the 1814 voyage of Allan Cunningham and James Bowie. They collected widely in South America, Australia and the Cape of Good Hope.

Due to Banks's patronage of such exploratory trips, numerous new plants were introduced into Britain. His advice on which vegetables, grains, herbs and fruits were best suited to the Australian climate was of immense value to the early settlers of that country. Banks felt strongly that Kew should become 'a great botanical exchange house for the empire' and he encouraged the free exchange of knowledge, as well as seeds, with botanical gardens in Ceylon (Sri Lanka), India and Caribbean islands such as Jamaica. He placed great emphasis on plants that had an economic value. For instance, in Tahiti he had encountered the edible breadfruit *Artocarpus altilis* and was later instrumental in its introduction to the West Indies in an attempt to feed the starving slave population.

Rosa banksiae is often referred to as Lady Banks's rose; it was named in honour of Sir Joseph Banks's wife and was collected by William Kerr on his 1803 expedition to China.

CARL LINNAEUS SAID OF SIR JOSEPH BANKS THAT HE WAS *VIR SINE PARI—A 'MAN WITHOUT EQUAL'*.

biflorus *BY-flo-rus*
biflora, biflorum
With twin flowers, as in *Geranium biflorum*

bifolius *by-FOH-lee-us*
bifolia, bifolium
With twin leaves, as in *Scilla bifolia*

bifurcatus *by-fur-KAH-tus*
bifurcata, bifurcatum
Divided into equal stems or branches, as in *Platycerium bifurcatum*

bignonioides *big-non-YOY-deez*
Resembling crossvine (*Bignonia*), as in *Catalpa bignonioides*

bijugus *bih-JOO-gus*
bijuga, bijugum
Two pairs joined together, as in *Pelargonium bijugum*

bilobatus *by-low-BAH-tus*
bilobata, bilobatum
bilobus *by-LOW-bus*
biloba, bilobum
With two lobes, as in *Ginkgo biloba*

bipinnatus *by-pin-NAH-tus*
bipinnata, bipinnatum
A leaf that is doubly pinnate, as in *Cosmos bipinnatus*

biserratus *by-ser-AH-tus*
biserrata, biserratum
A leaf that is double-toothed, as in *Nephrolepis biserrata*

biternatus *by-ter-NAH-tus*
biternata, biternatum
A leaf that is twice ternate, as in *Actaea biternata*

bituminosus *by-tu-min-OH-sus*
bituminosa, bituminosum
Like bitumen, sticky, as in *Bituminaria bituminosa*

bivalvis *by-VAL-vis*
bivalvis, bivalve
With two valves, as in *Ipheion bivalve*

blandus *BLAN-dus*
blanda, blandum
Mild or charming, as in *Anemone blanda*

blepharophyllus *blef-ar-oh-FIL-us*
blepharophylla, blepharophyllum
With leaves that are fringed like eyelashes, as in *Arabis blepharophylla*

bodinieri *boh-din-ee-ER-ee*
Named after Émile-Marie Bodinier (1842–1901), French missionary who collected plants in China, as in *Callicarpa bodinieri*

bodnantense *bod-nan-TEN-see*
Named after Bodnant Gardens, Wales, as in *Viburnum* × *bodnantense*

bonariensis *bon-ar-ee-EN-sis*
bonariensis, bonariense
From Buenos Aires, as in *Verbena bonariensis*

bonus *BOW-nus*
bona, bonum
In compound words, good, as in *Chenopodium bonus-henricus*, good
(King) Henry

borbonicus *bor-BON-ih-kus*
borbonica, borbonicum
Connected with Réunion Island in the Indian Ocean, formerly
known as Île Bourbon. Also refers to the French Bourbon kings, as in
Watsonia borbonica

borealis *bor-ee-AH-lis*
borealis, boreale
Northern, as in *Erigeron borealis*

borinquenus *bor-in-KAH-nus*
borinquena, borinquenum
From Borinquen, local name for Puerto Rico, as in *Roystonea
borinquena*

borneensis *bor-nee-EN-sis*
borneensis, borneense
From Borneo, as in *Gaultheria borneensis*

botryoides *bot-ROY-deez*
Resembling a bunch of grapes, as in *Muscari botryoides*

bowdenii *bow-DEN-ee-eye*
Named after plantsman Athelstan Cornish-Bowden (1871–1942), as
in *Nerine bowdenii*

brachiatus *brak-ee-AH-tus*
brachiata, brachiatum
With branches at right angles; like arms, as in *Clematis brachiata*

brachy-
Used in compound words to denote short

brachybotrys *brak-ee-BOT-rees*
With short clusters, as in *Wisteria brachybotrys*

brachycerus *brak-ee-SER-us*
brachycera, brachycerum
With short horns, as in *Gaylussacia brachycera*

brachypetalus *brak-ee-PET-uh-lus*
brachypetala, brachypetalum
With short petals, as in *Cerastium brachypetalum*

brachyphyllus *brak-ee-FIL-us*
brachyphylla, brachyphyllum
With short leaves, as in *Colchicum brachyphyllum*

bracteatus *brak-tee-AH-tus*
bracteata, bracteatum
bracteosus *brak-tee-OO-tus*
bracteosa, bracteosum
bractescens *brak-TES-senz*
With bracts, as in *Veltheimia bracteata*

brasilianus *bra-sill-ee-AHN-us*
brasiliana, brasilianum
brasiliensis *bra-sill-ee-EN-sis*
brasiliensis, brasiliense
From or of Brazil, as in *Begonia brasiliensis*

brevifolius *brev-ee-FOH-lee-us*
brevifolia, brevifolium
With short leaves, as in *Gladiolus brevifolius*

brevipedunculatus *brev-ee-ped-un-kew-LAH-tus*
brevipedunculata, brevipedunculatum
With a short flower stalk, as in *Olearia brevipedunculata*

brevis *BREV-is*
brevis, breve
Short, as in *Androsace brevis*

Rosa bracteata,
Macartney rose

Cotoneaster bullatus,
hollyberry cotoneaster

breviscapus *brev-ee-SKAY-pus*
breviscapa, breviscapum
With a short scape, as in *Lupinus breviscapus*

bromoides *brom-OY-deez*
Resembling brome grass (*Bromus*), as in *Stipa bromoides*

bronchialis *bron-kee-AL-lis*
bronchialis, bronchiale
Used in the past as a treatment for bronchitis, as in
Saxifraga bronchialis

brunneus *BROO-nee-us*
brunnea, brunneum
Deep brown, as in *Coprosma brunnea*

bryoides *bri-ROY-deez*
Resembling moss, as in *Dionysia bryoides*

buckleyi *BUK-lee-eye*
For commemorands named Buckley, such as William Buckley,
American geologist, as in *Schlumbergera × buckleyi*

bufonius *buf-OH-nee-us*
bufonia, bufonium
Relating to toads; grows in damp places, as in *Juncus bufonius*

bulbiferus *bulb-IH-fer-us*
bulbifera, bulbiferum
bulbiliferus *bulb-il-IH-fer-us*
bulbilifera, bulbiliferum
With bulbs, often referring to bulbils, as in *Lachenalia bulbifera*

bulbocodium *bulb-oh-KOD-ee-um*
With a woolly bulb, as in *Narcissus bulbocodium*

bulbosus *bul-BOH-sus*
bulbosa, bulbosum
A bulbous, swollen stem that grows underground; resembling
a bulb, as in *Ranunculus bulbosus*

bulgaricus *bul-GAR-ih-kus*
bulgarica, bulgaricum
Connected with Bulgaria, as in *Cerastium bulgaricum*

bullatus *bul-LAH-tus*
bullata, bullatum
With blistered or puckered leaves, as in *Cotoneaster bullatus*

bulleyanus *bul-ee-YAH-nus*
bulleyana, bulleyanum
bulleyi *bul-ee-YAH-eye*
Named after Arthur Bulley (1861–1942), founder of Ness Botanic
Gardens, Cheshire, England, as in *Primula bulleyana*

bungeanus *bun-jee-AH-nus*
bungeana, bungeanum
bungei *bun-jee-eye*
Named after Dr Alexander von Bunge (1803–1890), Russian
botanist, as in *Pinus bungeana*

burkwoodii *berk-WOOD-ee-eye*
Named after brothers Arthur and Albert Burkwood, 19th-century
hybridisers, as in *Viburnum × burkwoodii*

buxifolius *buks-ih-FOH-lee-us*
buxifolia, buxifolium
With leaves like box (*Buxus*), as in *Cantua buxifolia*

byzantinus *biz-an-TEE-nus*
byzantina, byzantinum
Connected with Istanbul, Turkey, as in *Colchicum byzantinum*

C

cacaliifolius *ka-KAY-see-eye-FOH-lee-us*
cacaliifolia, cacaliifolium
With leaves like *Cacalia*, as in *Salvia cacaliifolia*

cachemiricus *kash-MI-rih-kus*
cachemirica, cachemiricum
Connected with Kashmir, as in *Gentiana cachemirica*

cadierei *kad-ee-AIR-eye*
Named after R.P. Cadière, 20th-century plant collector in
Vietnam, as in *Pilea cadierei*

cadmicus *KAD-mih-kus*
cadmica, cadmicum
Metallic; like tin, as in *Ranunculus cadmicus*

caerulescens *see-roo-LES-enz*
Turning blue, as in *Euphorbia caerulescens*

caeruleus *see-ROO-lee-us*
caerulea, caeruleum
Dark blue, as in *Passiflora caerulea*

caesius *KESS-ee-us*
caesia, caesium
Bluish grey, as in *Allium caesium*

caespitosus *kess-pi-TOH-sus*
caespitosa, caespitosum
Growing in a dense clump, as in *Eschscholzia caespitosa*

caffer *KAF-er*
caffra, caffrum
caffrorum *kaf-ROR-um*
Connected with South Africa, as in *Erica caffra*

calabricus *ka-LA-brih-kus*
calabrica, calabricum
Connected with the Calabria region of Italy, as in
Thalictrum calabricum

calamagrostis *ka-la-mo-GROSS-tis*
From the Greek for reed-grass, as in *Stipa calamagrostis*

calamus *KAL-uh-mus*
From the Greek for reed, as in *Acorus calamus*

calandrinioides *ka-lan-DREEN-ee-oy-deez*
Resembling *Calandrinia*, as in *Ranunculus calandrinioides*

calcaratus *kal-ka-RAH-tus*
calcarata, calcaratum
With spurs, as in *Viola calcarata*

calcareus *kal-KAH-ree-us*
calcarea, calcareum
Relating to lime, as in *Titanopsis calcarea*

LATIN IN ACTION

Acorus calamus has numerous common names,
among them beewort, bitter pepper root, myrtle
sedge, pine root and sweet cane. *Calamus* refers to the
plant's reed-like leaves. For culinary use, its rhizome
can be dried and ground then used as a substitute for
cinnamon or ginger.

Acorus calamus,
sweet flag

calendulaceus *kal-en-dew-LAY-see-us*
calendulacea, calendulaceum
The colour of the yellow-flowered marigold (*Calendula officinalis*), as in *Rhododendron calendulaceum*

californicus *kal-ih-FOR-nih-kus*
californica, californicum
Connected with California, USA, as in *Zauschneria californica*

calleryanus *kal-lee-ree-AH-nus*
calleryana, calleryanum
Named after Joseph-Marie Callery (1810–1862), 19th-century French missionary who plant-hunted in France, as in *Pyrus calleryana*

callianthus *kal-lee-AN-thus*
calliantha, callianthum
With beautiful flowers, as in *Berberis calliantha*

callicarpus *kal-ee-KAR-pus*
callicarpa, callicarpum
With beautiful fruit, as in *Sambucus callicarpa*

callizonus *kal-ih-ZOH-nus*
callizona, callizonum
With beautiful bands or zones, as in *Dianthus callizonus*

callosus *kal-OH-sus*
callosa, callosum
With thick skin; with calluses, as in *Saxifraga callosa*

calophyllus *kal-ee-FIL-us*
calophylla, calophyllum
With beautiful leaves, as in *Dracocephalum calophyllum*

calvus *KAL-vus*
calva, calvum
Without hair; naked, as in *Viburnum calvum*

calycinus *ka-lih-KEE-nus*
calycina, calycinum
Like a calyx, as in *Halimium calycinum*

calyptratus *kal-lip-TRA-tus*
calyptrata, calyptratum
With a calyptra, a cap-like covering of a flower or fruit, as in *Podalyria calyptrata*

cambricus *KAM-brih-kus*
cambrica, cambricum
Connected with Wales, as in *Meconopsis cambrica*

campanularius *kam-pan-yoo-LAH-ri-us*
campanularia, campanularium
With bell-shaped flowers, as in *Phacelia campanularia*

campanulatus *kam-pan-yoo-LAH-tus*
campanulata, campanulatum
In the shape of a bell, as in *Enkianthus campanulatus*

campbellii *kam-BEL-ee-eye*
Named after Dr Archibald Campbell (1805–1874), Superintendent of Darjeeling, who accompanied Hooker to the Himalayas, as in *Magnolia campbellii*

campestris *kam-PES-tris*
campestris, campestre
Of fields or open plains, as in *Acer campestre*

camphoratus *kam-for-AH-tus*
camphorata, camphoratum
camphora *kam-for-AH*
Like camphor, as in *Thymus camphoratus*

campylocarpus *kam-plo-KAR-pus*
campylocarpa, campylocarpum
With curved fruit, as in *Rhododendron campylocarpum*

camtschatcensis *kam-shat-KEN-sis*
camtschatcensis, camtschatcense
camtschaticus *kam-SHAY-tih-kus*
camtschatica, camtschaticum
From or of the Kamchatka peninsula, Russia, as in *Lysichiton camtschatcensis*

canadensis *ka-na-DEN-sis*
canadensis, canadense
From Canada, though once also applied to north-eastern parts of the USA, as in *Cornus canadensis*

canaliculatus *kan-uh-lik-yoo-LAH-tus*
canaliculata, canaliculatum
With channels or grooves, as in *Erica canaliculata*

canariensis *kuh-nair-ee-EN-sis*
canariensis, canariense
From the Canary Islands, Spain, as in *Phoenix canariensis*

canbyi *KAN-bee-eye*
Named after William Marriott Canby (1831–1904), American botanist, as in *Quercus canbyi*

cancellatus *kan-sell-AH-tus*
cancellata, cancellatum
With cross bars, as in *Phlomis cancellata*

candelabrum *kan-del-AH-brum*
Branched like a candelabra as in *Salvia candelabrum*

candicans *KAN-dee-kanz*
candidus *KAN-dee-dus*
candida, candidum
Shining white, as in *Echium candicans*

canescens *kan-ESS-kenz*
With off-white or grey hairs, as in *Populus × canescens*

caninus *kay-NEE-nus*
canina, caninum
Relating to dogs, often meaning inferior, as in *Rosa canina*

cannabinus *kan-na-BEE-nus*
cannabina, cannabinum
Like hemp (*Cannabis*), as in *Eupatorium cannabinum*

cantabricus *kan-TAB-rih-kus*
cantabrica, cantabricum
Connected with the Cantabria region of Spain, as in
Narcissus cantabricus

canus *kan-nus*
cana, canum
Off-white; ash-coloured, as in *Calceolaria cana*

capensis *ka-PEN-sis*
capensis, capense
From the Cape of Good Hope, South Africa, as in
Phygelius capensis

capillaris *kap-ill-AH-ris*
capillaris, capillare
Very slender, like fine hair, as in *Tillandsia capillaris*

capillatus *kap-ill-AH-tus*
capillata, capillatum
With fine hairs, as in *Stipa capillata*

capillifolius *kap-ill-ih-FOH-lee-us*
capillifolia, capillifolium
With hairy leaves, as in *Eupatorium capillifolium*

capilliformis *kap-il-ih-FOR-mis*
capilliformis, capilliforme
Like hair, as in *Carex capilliformis*

capillipes *cap-ILL-ih-peez*
With slender feet, as in *Acer capillipes*

capillus-veneris *KAP-il-is VEN-er-is*
Venus's hair, as in *Adiantum capillus-veneris*

capitatus *kap-ih-TAH-tus*
capitata, capitatum
Flowers, fruit or whole plant growing in a dense head,
as in *Cornus capitata*

The beautiful and very fragrant Madonna lily is
frequently shown in religious paintings as an attribute
of the Virgin Mary, its white flowers a symbol of her
purity. The epithet *candidum* means shining white, a
particularly apt description of these blooms.

Lilium candidum,
Madonna lily

capitellatus *kap-ih-tel-AH-tus*
capitellata, capitellatum
capitellus *kap-ih-TELL-us*
capitella, capitellum
capitulatus *kap-ih-tu-LAH-tus*
capitulata, capitulatum
With a small head, as in *Primula capitellata*

cappadocicus *kap-puh-doh-SIH-kus*
cappadocica, cappadocicum
Connected with the ancient province of Cappadocia,
Asia Minor, as in *Omphalodes cappadocica*

capreolatus *kap-ree-oh-LAH-tus*
capreolata, capreolatum
With tendrils, as in *Bignonia capreolata*

capreus *KAP-ray-us*
caprea, capreum
Relating to goats, as in *Salix caprea*

Lobelia cardinalis,
cardinal flower

capricornis *kap-ree-KOR-nis*
capricornis, capricorne
Of or below the Tropic of Capricorn in the Southern Hemisphere;
shaped like a goat's horn, as in *Astrophytum capricorne*

caprifolius *kap-rih-FOH-lee-us*
caprifolia, caprifolium
With leaves having some characteristic of goats, as in *Lonicera
caprifolium*

capsularis *kap-SYOO-lah-ris*
capsularis, capsulare
With capsules, as in *Corchorus capsularis*

caracasanus *kar-ah-ka-SAH-nus*
caracasana, caracasanum
Connected with Caracas, Venezuela, as in *Serjania caracasana*

cardinalis *kar-dih-NAH-lis*
cardinalis, cardinale
Bright scarlet; cardinal red, as in *Lobelia cardinalis*

cardiopetalus *kar-dih-oh-PET-uh-lus*
cardiopetala, cardiopetalum
With heart-shaped petals, as in *Silene cardiopetala*

carduaceus *kard-yoo-AY-see-us*
carduacea, carduaceum
Like a thistle, as in *Salvia carduacea*

cardunculus *kar-DUNK-yoo-lus*
carduncula, cardunculum
Like a small thistle, as in *Cynara cardunculus*

caribaeus *kuh-RIB-ee-us*
caribaea, caribaeum
Connected with the Caribbean, as in *Pinus caribaea*

caricinus *kar-ih-KEE-nus*
caricina, caricinum
caricosus *kar-ee-KOH-sus*
caricosa, caricosum
Like sedge (*Carex*), as in *Dichanthium caricosum*

carinatus *kar-IN-uh-tus*
carinata, carinatum
cariniferus *Kar-in-IH-fer-us*
carinifera, cariniferum
With a keel, as in *Allium carinatum*

carinthiacus *kar-in-thee-AH-kus*
carinthiaca, carinthiacum
Connected with the Carinthia region of Austria, as in
Wulfenia carinthiaca

carlesii *KARLS-ee-eye*
Named after William Richard Carles (1848–1929) of the British
consular service in China, who collected plants in Korea, as in
Viburnum carlesii

carminatus *kar-MIN-uh-tus*
carminata, carminatum
carmineus *kar-MIN-ee-us*
carminea, carmineum
Carmine; bright crimson, as in *Metrosideros carminea*

carneus *KAR-nee-us*
carnea, carneum
Flesh colour; deep pink, as in *Androsace carnea*

carnicus *KAR-nih-kus*
carnica, carnicum
Like flesh, as in *Campanula carnica*

carniolicus *kar-nee-OH-lih-kus*
carniolica, carniolicum
Connected with the historical region of Carniola, now in Slovenia,
as in *Centaurea carniolica*

carnosulus *karn-OH-syoo-lus*
carnosula, carnosulum
Rather fleshy, as in *Hebe carnosula*

carnosus *kar-NOH-sus*
carnosa, carnosum
Fleshy, as in *Hoya carnosa*

carolinianus *kair-oh-lin-ee-AH-nus*
caroliniana, carolinianum
carolinensis *kair-oh-lin-ee-EN-sis*
carolinensis, carolinense
carolinus *kar-oh-LEE-nus*
carolina, carolinum
From or of North or South Carolina, USA, as in *Halesia carolina*

carota *kar-OH-tuh*
Carrot, as in *Daucus carota*

Hoya carnosa,
wax plant

carpaticus *kar-PAT-ih-kus*
carpatica, carpaticum
Connected with the Carpathian Mountains, as in
Campanula carpatica

carpinifolius *kar-pine-ih-FOH-lee-us*
carpinifolia, carpinifolium
With leaves like hornbeam (*Carpinus*), as in *Zelkova*
carpinifolia

carthusianorum *kar-thoo-see-an-OR-um*
Of Grande Chartreuse, Carthusian monastery near Grenoble,
France, as in *Dianthus carthusianorum*

cartilagineus *kart-ill-uh-GIN-ee-us*
cartilaginea, cartilagineum
Like cartilage, as in *Blechnum cartilagineum*

cartwrightianus *kart-RITE-ee-AH-nus*
Named after John Cartwright, 19th-century British consul to
Constantinople, as in *Crocus cartwrightianus*

caryophyllus *kar-ee-oh-FIL-us*
caryophylla, caryophyllum
Walnut-leaved (from Greek *karya*); likened to clove for their smell,
and thence to clove pink, as in *Dianthus caryophyllus*

Also known as wake robin and trinity flower, *Trillium catesbyi* was named after Mark Catesby (1682–1749), an English naturalist and painter. He wrote and illustrated the *Natural History of Carolina, Florida and the Bahama Islands*, an account of the regions' flora and fauna.

Trillium catesbyi, wood lily

caryopteridifolius *kar-ee-op-ter-id-ih-FOH-lee-us*
caryopteridifolia, caryopteridifolium
With leaves like *Caryopteris*, as in *Buddleja caryopteridifolia*

caryotideus *kar-ee-oh-TID-ee-us*
caryotidea, caryotideum
Like fishtail palms (*Caryota*), as in *Cyrtomium caryotideum*

cashmerianus *kash-meer-ee-AH-nus*
cashmeriana, cashmerianum
cashmirianus *kash-meer-ee-AH-nus*
cashmiriana, cashmirianum
cashmiriensis *kash-meer-ee-EN-sis*
cashmiriensis, cashmiriense
From or of Kashmir, as in *Cupressus cashmeriana*

caspicus *KAS-pih-kus*
caspica, caspicum
caspius *KAS-pee-us*
caspia, caspium
Connected with the Caspian Sea, as in *Ferula caspica*

catalpifolius *ka-tal-pih-FOH-lee-us*
catalpifolia, catalpifolium
With leaves like *Catalpa*, as in *Paulownia catalpifolia*

cataria *kat-AR-ee-uh*
Relating to cats, as in *Nepeta cataria*

catarractae *kat-uh-RAK-tay*
Of waterfalls, as in *Parahebe catarractae*

catawbiensis *ka-taw-bee-EN-sis*
catawbiensis, catawbiense
From the Catawba River, North Carolina, USA, as in *Rhododendron catawbiense*

catesbyi *KAYTS-bee-eye*
Named after Mark Catesby (1682–1749), English naturalist, as in *Sarracenia* × *catesbyi*

catharticus *kat-AR-tih-kus*
carthartica, catharticum
Cathartic; purgative, as in *Rhamnus cathartica*

cathayanus *kat-ay-YAH-nus*
cathayana, cathayanum
cathayensis *kat-ay-YEN-sis*
cathayensis, cathayense
From or of China, as in *Cardiocrinum cathayana*

caucasicus *kaw-KAS-ih-kus*
caucasica, caucasicum
Connected with the Caucasus, as in *Symphytum caucasicum*

caudatus *kaw-DAH-tus*
caudata, caudatum
With a tail, as in *Asarum caudatum*

caulescens *kawl-ESS-kenz*
With a stem, as in *Kniphofia caulescens*

cauliflorus *kaw-lih-FLOR-us*
cauliflora, cauliflorum
With flowers on the stem or trunk, as in *Saraca cauliflora*

causticus *KAWS-tih-kus*
caustica, causticum
With a caustic or burning taste, as in *Lithraea caustica*

cauticola *kaw-TIH-koh-luh*
Growing on cliffs, as in *Sedum cauticola*

cautleyoides *kawt-ley-OY-deez*
Resembling *Cautleya*, as in *Roscoea cautleyoides*

cavus *KA-vus*
cava, cavum
Hollow, as in *Corydalis cava*

cebennensis *kae-yen-EN-sis*
cebennensis, cebennense
From Cévennes, France, as in *Saxifraga cebennensis*

celastrinus *seh-lass-TREE-nus*
celastrina, celastrinum
Like the staff vine (*Celastrus*), as in *Azara celastrina*

centifolius *sen-tih-FOH-lee-us*
centifolia, centifolium
With many leaves; with a hundred leaves, as in *Rosa × centifolia*

centralis *sen-tr-AH-lis*
centralis, centrale
Central (for instance in distribution), as in *Diplocaulobium centrale*

centranthifolius *sen-tran-thih-FOH-lee-us*
centranthifolia, centranthifolium
With leaves like valerian (*Centranthus*), as in
Penstemon centranthifolius

cepa *KEP-uh*
The Roman name for an onion, as in *Allium cepa*

cephalonicus *kef-al-OH-nih-kus*
cephalonica, cephalonicum
Connected with Cephalonia, Greece, as in
Abies cephalonica

cephalotes *sef-ah-LOH-tees*
Like a small head, as in *Gypsophila cephalotes*

ceraceus *ke-ra-KEE-us*
ceracea, ceraceum
With a waxy texture, as in *Wahlenbergia ceracea*

ceramicus *ke-RA-mih-kus*
ceramica, ceramicum
Like pottery, as in *Rhopaloblaste ceramica*

cerasiferus *ke-ra-SIH-fer-us*
cerasifera, cerasiferum
A plant that bears cherries or cherry-like fruit, as in *Prunus cerasifera*

cerasiformis *see-ras-if-FOR-mis*
cerasiformis, cerasiforme
Shaped like a cherry, as in *Oemleria cerasiformis*

cerasinus *ker-ras-EE-nus*
cerasina, cerasinum
Cherry-red, as in *Rhododendron cerasinum*

cerastiodes *ker-ras-tee-OY-deez*
cerastioides
Resembling mouse-ear chickweed (*Cerastium*), as in
Arenaria cerastioides

cerasus *KER-uh-sus*
Latin for cherry, as in *Prunus cerasus*

Rosa × centifolia 'Foliacée',
cabbage rose

cercidifolius *ser-uh-sid-ih-FOH-lee-us*
cercidifolia, cercidifolium
With leaves like redbud tree (*Cercis*), as in *Disanthus cercidifolius*

cerealis *ser-ee-AH-lis*
cerealis, cereale
Relating to agriculture, derived from Ceres, the goddess of farming, as in *Secale cereale*

cerefolius *ker-ee-FOH-lee-us*
cerefolia, cerefolium
With waxy leaves, as in *Anthriscus cerefolium*

cereus *ker-REE-us*
cerea, cereum
cerinus *ker-REE-nus*
cerina, cerinum
Waxy, as in *Ribes cereum*

Poa chaixii,
broad-leaved meadow grass

ceriferus *ker-IH-fer-us*
cerifera, ceriferum
Producing wax, as in *Morella cerifera*

cerinthoides *ser-in-THOY-deez*
Resembling honeywort (*Cerinthe*), as in *Tradescantia cerinthoides*

cernuus *SER-new-us*
cernua, cernuum
Drooping or nodding, as in *Enkianthus cernuus*

ceterach *KET-er-ak*
Derived from an Arabic word applied to spleenworts (*Asplenium*), as in *Asplenium ceterach*

chaixii *kay-IKX-ee-eye*
Named after Dominique Chaix (1730–99), French botanist, as in *Verbascum chaixii*

chalcedonicus *kalk-ee-DON-ih-kus*
chalcedonica, chalcedonicum
Connected with Chalcedon, the ancient name for a district of Istanbul, Turkey, as in *Lychnis chalcedonica*

chamaebuxus *kam-ay-BUKS-us*
Dwarf boxwood, as in *Polygala chamaebuxus*

chamaecyparissus *kam-ee-ky-pah-RIS-us*
Like *Chamaecyparis*, as in *Santolina chamaecyparissus*

chamaedrifolius *kam-ee-drih-FOH-lee-us*
chamaedrifolia, chamaedrifolium
chamaedryfolius
chamaedryfolia, chamaedryfolium
With leaves like *Chamaedrys*, as in *Aloysia chamaedrifolia* (note: *Chamaedrys* is now listed under *Teucrium*).

chantrieri *shon-tree-ER-ee*
Named after the French nursery Chantrier Frères, as in *Tacca chantrieri*

charianthus *kar-ee-AN-thus*
chariantha, charianthum
With elegant flowers, as in *Ceratostema charianthum*

chathamicus *chath-AM-ih-kus*
chathamica, chathamicum
Connected with the Chatham Islands, in the South Pacific, as in *Astelia chathamica*

cheilanthus *kay-LAN-thus*
cheilantha, cheilanthum
With flowers that have a lip, as in *Delphinium cheilanthum*

cheiri *kye-EE-ee*

Perhaps from the Greek work *cheir*, a hand, as in *Erysimum cheiri*

chelidonioides *kye-li-don-OY-deez*

Resembling greater celandine (*Chelidonium*), as in *Calceolaria chelidonioides*

chilensis *chil-ee-EN-sis*

chilensis, chilense

From Chile, as in *Blechnum chilense*

chiloensis *kye-loh-EN-sis*

chiloensis, chiloense

From the island of Chiloe, Chile, as in *Fragaria chiloensis*

chinensis *CHI-nen-sis*

chinensis, chinense

From China, as in *Stachyurus chinensis*

chionanthus *kee-on-AN-thus*

chionantha, chionanthum

With snow-white flowers, as in *Primula chionantha*

chloranthus *klor-ah-AN-thus*

chlorantha, chloranthum

With green flowers, as in *Fritillaria chlorantha*

chlorochilon *klor-oh-KY-lon*

With a green lip, as in *Cycnoches chlorochilon*

chloropetalus *klo-ro-PET-al-lus*

chloropetala, chloropetalum

With green petals, as in *Trillium chloropetalum*

chrysanthus *kris-AN-thus*

chrysantha, chrysanthum

With golden flowers, as in *Crocus chrysanthus*

chryseus *KRIS-ee-us*

chrysea, chryseum

Golden, as in *Dendrobium chryseum*

chrysocarpus *kris-oh-KAR-pus*

chrysocarpa, chrysocarpum

With golden fruit, as in *Crataegus chrysocarpa*

chrysocomus *kris-oh-KOH-mus*

chrysocoma, chrysocomum

With golden hairs, as in *Clematis chrysocoma*

Cheiri (as in *Erysimum cheiri*) is a good example of an epithet whose origins have become obscure. According to one theory, it derives from the Greek *cheir*, hand, in reference to the use of wallflowers in bouquets. Alternatively, some sources suggest a connection with an Arabic word for red, though the species is not typically associated with red flowers.

Erysimum cheiri, wallflower

chrysographes *kris-oh-GRAF-ees*

With gold markings, as in *Iris chrysographes*

chrysolepis *kris-SOL-ep-is*

chrysolepis, chrysolepe

With golden scales, as in *Quercus chrysolepis*

chrysoleucus *kris-roh-LEW-kus*

chrysoleuca, chrysoleucum

Gold and white, as in *Hedychium chrysoleucum*

chrysophyllus *kris-oh-FIL-us*

chrysophylla, chrysophyllum

With golden leaves, as in *Phlomis chrysophylla*

MERIWETHER LEWIS

(1774–1809)

WILLIAM CLARK

(1770–1838)

Meriwether Lewis met the experienced explorer William Clark while serving as a lieutenant in the American Army. Both men had been active in the Northwest Campaign fighting against the British and Native Americans. Lewis became President Thomas Jefferson's secretary-aide and was appointed by Jefferson to lead the 1804 Corps of Discovery expedition to find a passable route from the Mississippi River to the Pacific. Well aware of Clark's leadership qualities, Lewis chose his former military colleague as the co-leader of what he knew would prove to be a gruelling mission. Having studied botany at the University of Pennsylvania as a preparation for the expedition, Lewis became its official naturalist and Clark, after gaining expertise in astronomy and map-making, its cartographer. In a deal known as the Louisiana Purchase, the area Lewis and Clark were to map had recently been acquired by the Americans from the French and was more or less uncharted territory. The two men were commissioned by Jefferson to record the geography of the region, note routes and landmarks, collect data on weather conditions, study the geology, analyse soil types and observe the native flora and fauna.

The journey was not without hazard; the rivers had to be traversed by canoe and during their passage the Rocky Mountains were deep in snow. However, they finally reached the Pacific in the autumn of 1805. Along the trail they collected huge quantities of plant material and succeeded in identifying around 170 plants that were new to science. In the prairies they collected the wild rose *Rosa arkansana* and at the Missouri River's Great Falls they encountered a tree that David Douglas would later identify as the Douglas fir, *Pseudotsuga menziesii* (see p. 111). *Rhus aromatica*, the skunkbush sumac, was another of their finds. They noted the abundance of fruit that grew in the region, including the silver-leaf Indian breadroot *Pediomelum argophyllum*, and traded corn and squash with Native American tribes.

Lewis and Clark's ground-breaking expedition paved the way for later settlers, and they returned to Washington national heroes. Jefferson wrote of Lewis in the most glowing of terms:

These two men at the forefront of the Corps of Discovery expedition were of very different character: while Lewis (left) was known to be introverted by nature, Clark (right) was an extrovert.

*Of courage undaunted, possessing a firmness
and perseverance of purpose which nothing but
impossibilities could divert from its direction ... honest,
disinterested, liberal, of sound understanding and a
fidelity to truth so scrupulous that whatever he should
report would be as certain as if seen by ourselves.*

Lewis became the Governor of Louisiana Territory
(a post in which he was not an unqualified success),
but died in uncertain circumstances aged only 35. Clark
lived into his 60s and went on to become the Governor
of Missouri Territory and Superintendent of Indian
Affairs, gaining much respect. Subsequently, several
plants were named in honour of these intrepid explorers,
including *Lewisia rediviva* and *Clarkia unguiculata*.

Due to the extreme difficulty of much of the
terrain encountered on the expedition, many of the
specimens collected by Lewis and Clark were lost en
route. However, those that survived they gave to the
German botanist Frederick Pursh to record. Pursh
had arrived in America in 1799 and settled in
Philadelphia. He met William Bartram (see p. 98),
and while working for him as a curator and collector
he secured the patronage of the famous botanist
Benjamin Smith Barton, who funded his American

Philadelphus lewisii,
mock orange

This is one of the many plants collected by Lewis and Clark on their
exploratory overland expedition to the Pacific Ocean. Among the plants
named in honour of Clark is *Clarkia pulchella*, variously known as
deerhorn clarkia, ragged robin or pink fairies.

German botanist
Frederick Pursh recorded
many of the plants
collected by Lewis and
Clark, including this
Clarkia pulchella painted
in 1814.

travels collecting plants. Plants that bear Pursh's
name include *Frangula purshiana*, the bitterbush. In
1807, Lewis employed Pursh to catalogue the plant
specimens from his and Clark's expedition. However,
it was not until 1814 that Pursh first published in
Europe his *Flora Americae Septentrionalis; or, a
Systematic Arrangement and Description of the Plants
of North America*. This tome included more than
130 of the plants collected by Lewis and Clark and
also proposed 94 new names based on their findings,
40 of which are still in use today. Pursh's efforts to
record the findings of Lewis and Clark contributed
in no small way to their enduring legacy.

'[LEWIS HAD] A LUMINOUS AND
DISCRIMINATING INTELLECT.'
Thomas Jefferson (1743–1826)

chrysostoma *kris-oh-STO-muh*
With a golden mouth, as in *Lasthenia chrysostoma*

cicutarius *kik-u-tah-ree-us*
cicutaria, cicutarium
Like water hemlock (*Conium maculatum*, formerly *Cicuta*), as in *Erodium cicutarium*

ciliaris *sil-ee-AH-ris*
ciliaris, ciliare
ciliatus *sil-ee-ATE-us*
ciliata, ciliatum
With leaves and petals that are fringed with hairs, as in *Tropaeolum ciliatum*

cilicicus *kil-LEE-kih-kus*
cilicica, cilicicum
Connected with Lesser Armenia (formerly Cilicia), as in *Colchicum cilicicum*

ciliicalyx *kil-LEE-kal-ux*
With a fringed calyx, as in *Menziesia ciliicalyx*

ciliosus *sil-ee-OH-sus*
ciliosa, ciliosum
With a small fringe, as in *Sempervivum ciliosum*

cinctus *SINK-tus*
cincta, cinctum
With a girdle, as in *Angelica cincta*

cinerariifolius *sin-uh-rar-ee-ay-FOH-lee-us*
cinerariifolia, cinerariifolium
With leaves like *Cineraria*, as in *Tanacetum cinerariifolium*

cinerarius *sin-uh-RAH-ree-us*
cineraria, cinerarium
Ash-grey, as in *Centaurea cineraria*

cinerascens *sin-er-ASS-enz*
Turning to ash-grey, as in *Senecio cinerascens*

cinereus *sin-EER-ee-us*
cinerea, cinereum
The colour of ash, as in *Veronica cinerea*

cinnabarinus *sin-uh-bar-EE-nus*
cinnabarina, cinnabarinum
Vermilion, as in *Echinopsis cinnabarina*

cinnamomeus *sin-uh-MOH-mee-us*
cinnamomea, cinnamomeum
Cinnamon brown, as in *Osmunda cinnamomea*

cinnamomifolius *sin-nuh-mom-ih-FOH-lee-us*
cinnamomifolia, cinnamomifolium
With leaves like cinnamon (*Cinnamomum*), as in *Viburnum cinnamomifolium*

circinalis *kir-KIN-ah-lis*
circinalis, circinale
Coiled in form, as in *Cycas circinalis*

circum-
Used in compound words to denote around

cirratus *sir-RAH-tus*
cirrata, cirratum
cirrhosus *sir-ROH-sus*
cirrhosa, cirrhosum
With tendrils, as in *Clematis cirrhosa*

cissifolius *kiss-ih-FOH-lee-us*
cissifolia, cissifolium
With leaves like ivy (from the Greek *kissos*), as in *Acer cissifolium*

cistena *sis-TEE-nuh*
Of dwarf habit, from the Sioux word for baby, as in *Prunus × cistena*

citratus *sit-TRAH-tus*
citrata, citratum
Like *Citrus*, as in *Mentha citrata*

citrinus *sit-REE-nus*
citrina, citrinum
Lemon yellow or like *Citrus*, as in *Callistemon citrinus*

citriodorus *sit-ree-oh-DOR-us*
citriodora, citriodorum
With the scent of lemons, as in *Thymus citriodorus*

citrodora *sit-roh-DOR-uh*
With a lemon scent, as in *Aloysia citrodora*

cladocalyx *kla-do-KAL-iks*
From the Greek *klados*, a branch, referring to flowers borne on leafless branches, as in *Eucalyptus cladocalyx*

clandestinus *klan-des-TEE-nus*
clandestina, clandestinum
Hidden; concealed, as in *Lathraea clandestina*

clandonensis *klan-don-EN-sis*
From Clandon, England, as in *Caryopteris × clandonensis*

clarkei *KLAR-kee-eye*
Commemorates various noteworthy people with the surname Clarke, including Charles Baron Clarke (1832–1906), Superintendent of the Calcutta Botanic Gardens and former president of the Linnean Society, as in *Geranium clarkei*

clausus *KLAW-sus*
clausa, clausum
Closed; shut, as in *Pinus clausa*

clavatus *KLAV-ah-tus*
clavata, clavatum
Shaped like a club, as in *Calochortus clavatus*

claytonianus *klay-ton-ee-AH-nus*
claytoniana, claytonianum
Named after John Clayton (1694–1773), plant collector in Virginia, USA, as in *Osmunda claytoniana*

clematideus *klem-AH-tee-dus*
clematidea, clematideum
Like *Clematis*, as in *Agdestis clematidea*

clethroides *klee-THROY-deez*
Resembling white alder (*Clethra*), as in *Lysimachia clethroides*

clevelandii *kleev-LAN-dee-eye*
Named after Daniel Cleveland, 19th-century American collector and fern expert, as in *Bloomeria clevelandii*

clusianus *kloo-zee-AH-nus*
clusiana, clusianum
Named after Charles de l'Écluse (1526–1609), Flemish botanist, as in *Tulipa clusiana*

clypeatus *klye-pee-AH-tus*
clypeata, clypeatum
Like the round Roman shield, as in *Fibigia clypeata*

clypeolatus *klye-pee-OH-la-tus*
clypeolata, clypeolatum
Shaped rather like a shield, as in *Achillea clypeolata*

cneorum *suh-NOR-um*
From the Greek for a small shrub like olive, possibly a kind of *Daphne*; as in *Convolvulus cneorum*

coarctatus *koh-ARK-tah-tus*
coarctata, coarctatum
Pressed or crowded together, as in *Achillea coarctata*

Tulipa clusiana,
lady tulip

cocciferus *koh-KIH-fer-us*
coccifera, cocciferum
coccigerus *koh-KEE-ger-us*
coccigera, coccigerum
Producing berries, as in *Eucalyptus coccifera*

coccineus *kok-SIN-ee-us*
coccinea, coccineum
Scarlet, as in *Musa coccinea*

cochlearis *kok-lee-AH-ris*
cochlearis, cochleare
Shaped like a spoon, as in *Saxifraga cochlearis*

cochleatus *kok-lee-AH-tus*
cochleata, cochleatum
Shaped like a spiral, as in *Lycaste cochleata*

cockburnianus *kok-burn-ee-AH-nus*
cockburniana, cockburnianum
Named after the family Cockburn, residents of China,
as in *Rubus cockburnianus*

coelestinus *koh-el-es-TEE-nus*
coelestina, coelestinum
coelestis *koh-el-ES-tis*
coelestis, coeleste
Sky-blue, as in *Phalocallis coelestis*

coeruleus *ko-er-OO-lee-us*
coerulea, coeruleum
Blue, as in *Satureja coerulea*

cognatus *kog-NAH-tus*
cognata, cognatum
Closely related to, as in *Acacia cognata*

colchicus *KOHL-chih-kus*
colchica, colchicum
Connected with the coastal region of the Black Sea, Georgia,
as in *Hedera colchica*

colensoi *co-len-SO-ee*
Named after Revd William Colenso (1811–99), New Zealand plant
collector, as in *Pittosporum colensoi*

collinus *kol-EE-nus*
collina, collinum
Relating to hills, as in *Geranium collinum*

colorans *kol-LOR-anz*
coloratus *kol-or-AH-tus*
colorata, coloratum
Coloured, as in *Silene colorata*

colubrinus *kol-oo-BREE-nus*
colubrina, colubrinum
Like a snake, as in *Opuntia colubrina*

columbarius *kol-um-BAH-ree-us*
columbaria, columbarium
Like a dove, as in *Scabiosa columbaria*

columbianus *kol-um-bee-AH-nus*
columbiana, columbianum
Connected with British Columbia, Canada, as in
Aconitum columbianum

columellaris *kol-um-EL-ah-ris*
columellaris, columellare
Relating to a small pillar or pedestal, as in
Callitris columellaris

columnaris *kol-um-nah-ris*
columnaris, columnare
In the shape of a column, as in *Eryngium columnare*

colvillei *koh-VIL-ee-eye*
Named after either Sir James William Colville (1801–1880),
Scottish lawyer and judge in Calcutta, or James Colville, 19th-
century nurseryman, as in *Gladiolus × colvillei* (after the latter)

comans *KO-manz*
comatus *kom-MAH-tus*
comata, comatum
Tufted, as in *Carex comans*

commixtus *kom-miks-tus*
commixta, commixtum
Mixed; mingled together, as in *Sorbus commixta*

communis *KOM-yoo-nis*
communis, commune
Growing in groups; common, as in *Myrtus communis*

commutatus *kom-yoo-TAH-tus*
commutata, commutatum
Changed, for instance when formerly included in another species, as
in *Papaver commutatum*

Pyrus communis,
pear

comosus *kom-OH-sus*
comosa, comosum
With tufts, as in *Eucomis comosa*

compactus *kom-PAK-tus*
compacta, compactum
Compact; dense, as in *Pleiospilos compactus*

complanatus *kom-plan-NAH-tus*
complanata, complanatum
Flat; level, as in *Lycopodium complanatum*

complexus *kom-PLEKS-us*
complexa, complexum
Complex; encircled; as in *Muehlenbeckia complexa*

complicatus *kom-plih-KAH-tus*
complicata, complicatum
Complicated; complex, as in *Adenocarpus complicatus*

compressus *kom-PRESS-us*
compressa, compressum
Compressed, flattened, as in *Conophytum compressum*

comptoniana *komp-toh-nee-AH-nuh*
For various people with the surname Compton, as in *Hardenbergia comptoniana*

concavus *kon-KAV-us*
concava, concavum
Hollowed out, as in *Conophytum concavum*

conchifolius *con-chee-FOH-lee-us*
conchifolia, conchifolium
With leaves shaped like sea shells, as in *Begonia conchifolia*

concinnus *KON-kin-us*
concinna, concinnum
With a neat or elegant form, as in *Parodia concinna*

concolor *KON-kol-or*
All the same colour, as in *Abies concolor*

condensatus *kon-den-SAH-tus*
condensata, condensatum
condensus *kon-DEN-sus*
condensa, condensum
Crowded together, as in *Alyssum condensatum*

confertiflorus *kon-fer-tih-FLOR-us*
confertiflora, confertiflorum
With flowers crowded together, as in *Salvia confertiflora*

confertus *KON-fer-tus*
conferta, confertum
Crowded together, as in *Polemonium confertum*

confusus *kon-FEW-sus*
confusa, confusum
Confused or uncertain, as in *Sarcococca confusa*

congestus *kon-JES-tus*
congesta, congestum
Congested or crowded together, as in *Aciphylla congesta*

conglomeratus *kon-glom-er-AH-tus*
conglomerata, conglomeratum
Crowded together, as in *Cyperus conglomeratus*

conicus *KON-ih-kus*
conica, conicum
In the shape of a cone, as in *Carex conica*

coniferus *koh-NIH-fer-us*
conifera, coniferum
With cones, as in *Magnolia conifera*

conjunctus *kon-JUNK-tus*
conjuncta, conjunctum
Joined, as in *Alchemilla conjuncta*

connatus *kon-NAH-tus*
connata, connatum
United; twin; opposite leaves joined together at the base, as in *Bidens connata*

conoideus *ko-NOY-dee-us*
conoidea, conoideum
Resembling a cone, as in *Silene conoidea*

conopseus *kon-OP-see-us*
conopsea, conopseum
Gnat-like, from Greek *konops*, as in *Gymnadenia conopsea*

consanguineus *kon-san-GWIN-ee-us*
consanguinea, consanguineum
Related, as in *Vaccinium consanguineum*

conspersus *kon-SPER-sus*
conspersa, conspersum
Scattered, as in *Primula conspersa*

conspicuus *kon-SPIK-yoo-us*
conspicua, conspicuum
Conspicuous, as in *Sinningia conspicua*

Silene conica,
striped corn catchfly

constrictus *kon-STRIK-tus*
constricta, constrictum
Constricted, as in *Yucca constricta*

contaminatus *kon-tam-in-AH-tus*
contaminata, contaminatum
Contaminated; defiled, as in *Lachenalia contaminata*

continentalis *kon-tin-en-TAH-lis*
continentalis, continentale
Continental, as in *Aralia continentalis*

contortus *kon-TOR-tus*
contorta, contortum
Twisted; contorted, as in *Pinus contorta*

contra-
Used in compound words to denote against

contractus *kon-TRAK-tus*
contracta, contractum
Contracted; drawn together, as in *Fargesia contracta*

controversus *kon-troh-VER-sus*
controversa, controversum
Controversial; doubtful, as in *Cornus contraversa*

convallarioides *kon-va-lar-ee-OY-deez*
Resembling lily-of-the-valley (*Convallaria*), as in *Speirantha convallarioides*

convolvulaceus *kon-vol-vu-la-SEE-us*
convolvulacea, convolvulaceum
Rather like *Convolvulus*, as in *Codonopsis convolvulacea*

conyzoides *kon-ny-ZOY-deez*
Resembling *Conyza*, as in *Ageratum conyzoides*

copallinus *kop-al-EE-nus*
copallina, copallinum
With gum or resin, as in *Rhus copallinum*

coralliflorus *kaw-lih-FLOR-us*
coralliflora, coralliflorum
With coral-red flowers, as in *Lampranthus coralliflorus*

corallinus *kor-al-LEE-nus*
corallina, corallinum
Coral-red, as in *Ilex corallina*

coralloides *kor-al-OY-deez*
Resembling coral, as in *Ozothamnus coralloides*

cordatus *kor-DAH-tus*
cordata, cordatum
In the shape of a heart, as in *Pontederia cordata*

cordifolius *kor-di-FOH-lee-us*
cordifolia, cordifolium
With heart-shaped leaves, as in *Crambe cordifolia*

cordiformis *kord-ih-FOR-mis*
cordiformis, cordiforme
Heart-shaped, as in *Carya cordiformis*

coreanus *kor-ee-AH-nus*
coreana, coreanum
Connected with Korea, as in *Hemerocallis coreana*

coriaceus *kor-ee-uh-KEE-us*
coriacea, coriaceum
Thick, tough and leathery, as in *Paeonia coriacea*

coriarius *kor-i-AH-ree-us*
coriaria, coriarium
Like leather, as in *Caesalpinia coriaria*

coridifolius *kor-id-ee-FOH-lee-us*
coridifolia, coridifolium
coriophyllus *kor-ee-uh-FIL-us*
coriophylla, coriophyllum
With leaves like *Coris*, as in *Erica corifolia*

corifolius *kor-ee-FOH-lee-us*
corifolia, corifolium
coriifolius *kor-ee-eye-FOH-lee-us*
coriifolia, coriifolium
With leathery leaves, as in *Erica corifolia*

corniculatus *korn-ee-ku-LAH-tus*
corniculata, corniculatum
With small horns, as in *Lotus corniculatus*

corniferus *korn-IH-fer-us*
conifera, coniferum
corniger *korn-ee-ger*
cornigera, cornigerum
With horns, as in *Coryphantha cornifera*

cornucopiae *korn-oo-KOP-ee-ay*
Of a cornucopia or horn of plenty, as in *Fedia cornucopiae*

cornutus *kor-NOO-tus*
cornuta, cornutum
With horns or shaped like a horn, as in *Viola cornuta*

corollatus *kor-uh-LAH-tus*
corollata, corollatum
Like a corolla, as in *Fuchsia corollata*

coronans *kor-OH-nanz*
coronatus *kor-oh-NAH-tus*
coronata, coronatum
Crowned, as in *Lychnis coronata*

coronarius *kor-oh-NAH-ree-us*
coronaria, coronarium
Used for garlands, as in *Anemone coronaria*

coronopifolius *koh-ron-oh-pih-FOH-lee-us*
coronopifolia, coronopifolium
With leaves like *Coronopus*, as in *Lobelia coronopifolia*

corrugatus *kor-yoo-GAH-tus*
corrugata, corrugatum
Corrugated; wrinkled, as in *Salvia corrugata*

Lotus corniculatus,
bird's foot trefoil

corsicus *KOR-sih-kus*
corsica, corsicum
Connected with Corsica, France, as in *Crocus corsicus*

cortusoides *kor-too-SOY-deez*
Resembling *Cortusa*, as in *Primula cortusoides*

corylifolius *kor-ee-lee-FOH-lee-us*
corylifolia, corylifolium
With leaves like hazelnut (*Corylus*), as in *Betula corylifolia*

corymbiferus *kor-im-BIH-fer-us*
corymbifera, corymbiferum
With a corymb, as in *Linum corymbiferum*

corymbiflorus *kor-im-BEE-flor-us*
corymbiflora, corymbiflorum
With flowers produced in a corymb, as in *Solanum corymbiflorum*

corymbosus *kor-rim-BOH-sus*
corymbosa, corymbosum
With corymbs, as in *Vaccinium corymbosum*

cosmophyllus *kor-mo-FIL-us*
cosmophylla, cosmophyllum
With leaves like *Cosmos*, as in *Eucalyptus cosmophylla*

costatus *kos-TAH-tus*
costata, costatum
With ribs, as in *Aglaonema costatum*

cotinifolius *kot-in-ih-FOH-lee-us*
cotinifolia, cotinifolium
With leaves like smoke tree (*Cotinus*), as in *Euphorbia cotinifolia*

cotyledon *kot-EE-lee-don*
A small cup (referring to the leaves), as in *Lewisia cotyledon*

coulteri *kol-TER-ee-eye*
Named after Dr Thomas Coulter (1793–1843), Irish botanist,
as in *Romneya coulteri*

coum *KOO-um*
Connected with Kos, Greece, as in *Cyclamen coum*

crassicaulis *krass-ih-KAW-lis*
crassicaulis, crassicaule
With a thick stem, as in *Begonia crassicaulis*

crassifolius *krass-ih-FOH-lee-us*
crassifolia, crassifolium
With thick leaves, as in *Pittosporum crassifolium*

LATIN IN ACTION

In acknowledgement of its Greek origins (it is
connected to the island of Kos) this lovely bulbous
plant is also commonly known as the eastern
cyclamen. Flowering in late winter and early spring,
it has deep pink or purple flowers set against green
or silver-grey foliage.

Cyclamen coum,
round-leaved cyclamen

---

crassipes *KRASS-ih-peez*
With thick feet or thick stems, as in *Quercus crassipes*

crassiusculus *krass-ih-US-kyoo-lus*
crassiuscula, crassiusculum
Quite thick, as in *Acacia crassiuscula*

crassus *KRASS-us*
crassa, crassum
Thick; fleshy, as in *Asarum crassum*

Plants: Their Shape and Form

When you are deciding what plant to put where, it is essential to know what shape, form and habit it will have once fully grown. Whether choosing a plant that is destined for the back of a border, the front of a bed or indeed a window box, it is of immense help for the gardener to know that a plant with *scandens* in its name likes to climb, while plants called *repens* have a creeping habit. Similarly, you should be aware that the perennial grass *Miscanthus sinensis* 'Strictus' grows into a tall upright clump and *Melica altissima* reaches well over a metre. (*Strictus, stricta, strictum*, meaning erect and *altissimus, altissima, altissimum* is very tall or tallest.)

It is also very important to know the eventual size that a plant will reach. When seeking something tall and stately, look for labels bearing *altus* (*alta, altum*, tall), *elatus* (*elata, elatum*, taller) and *elatior* or *excelsus* (*excelsa, excelsum*, taller). By comparison, short forms may be described as *brevis* (*brevis, breve*, short), *jejunus* (*jejuna, jejunum*, small), *minutus* (*minuta, minutum*, very small), *nanus* (*nana, nanum*, dwarf) or even *nanellus* (*nanella, nanellum*, very dwarf). These are just a few of the terms in botanical Latin that refer to scale, but do be aware that they may also relate to a particular part of the plant, or a comparison to the plant's close relatives, rather than its overall form. For instance, the prefix *macro-* means long or large but may simply describe a plant's large leaves, as in *macrophyllus* (*macrophylla, macrophyllum*), while the plant on which they grow might be relatively small. Likewise, *micro-*, small, can be applied to various elements such as small fruits, *microcarpus* (*microcarpa, microcarpum*), or small petals, *micropetalus* (*micropetala, micropetalum*). Some terms are more obvious – for instance, the rare *cyclops* and *titanus*, which mean huge or enormous!

Shape as well as size is often indicated by the name of a plant; again, these terms can refer either to the whole plant or just to a component part. For example, *arctuatus* (*arctuata, arctuatum*) means shaped like an arc or bow, *cruciatus* (*cruciata, cruciatum*) means in the shape of a cross, and *orbicularis* (*orbicularis, orbiculare*) means flat and round, like a disc. *Crenatus* (*crenata, crenatum*) means having scallop-shaped cuts or crenations, as in *Ranunculus crenatus*, an alpine with attractively crenated edges to its leaves. Similarly, the flowers of plants named *crenatiflorus* (*crenatiflora, crenatiflorum*) have flowers cut into wavy, rounded scallops, as in *Calceolaria crenatiflora*.

Cobaea scandens
cup and saucer plant

This plant can easily reach 3 m (10 ft) in a good summer.

Nardus stricta,
mat grass

The upright and erect stems of the aptly named *Nardus stricta* bear attractive yellow anthers that turn to white.

Rosa tomentosa,
harsh downy rose

Tomentosa means woolly or matted, hence the other common name for this rose, the whitewoolly rose and like *Pelargonium tomentosum* it has lovely soft leaves.

A surprisingly large number of terms relate to the woolly and hairy qualities of plants. Both the prefixes *eri-* and *lasi-* mean woolly. An *eriantherus* (*erianthera, eriantherum*) plant will have woolly anthers, while one with woolly spines is called *lasiacanthus* (*lasiacantha, lasiacanthum*). *Mollis*, (*mollis, molle*) means soft or with soft hairs and often refers to the soft or hairy leaves of many plants, like *Geranium molle* or *Alchemilla mollis*. In the case of *Astragalus mollissimus*, the reference is to the soft hairs of its flower stems. (*Mollissimus, mollissima* and *mollissimum* all mean very soft.)

Some descriptive terms are quite poetic; for instance, *nebulosus* means like a cloud (*nebulosa, nebulosum*) and *nubicola* means growing up in the clouds. However, this does not refer to the lofty height of the plant but to the fact that such plants as *Salvia nubicola* are found growing at high altitudes. Occasionally a botanical term can seem fanciful rather than helpful if it assumes a somewhat anthropomorphic tone. For instance, *superciliaris* (*superciliaris, superciliare*) means like an eyebrow – presumably one raised with a supercilious air, as in the orchid *Cypripedium × superciliare*!

crataegifolius *krah-tee-gi-FOH-lee-us*
crataegifolia, crataegifolium
With leaves like hawthorn (*Crataegus*), as in *Acer crataegifolium*

crenatiflorus *kren-at-ih-FLOR-us*
crenatiflora, crenatiflorum
With flowers cut into rounded scallops, as in *Calceolaria crenatiflora*

crenatus *kre-NAH-tus*
crenata, crenatum
Scalloped, crenate, as in *Ilex crenata*

crenulatus *kren-yoo-LAH-tus*
crenulata, crenulatum
Rather scalloped, as in *Boronia crenulata*

crepidatus *krep-id-AH-tus*
crepidata, crepidatum
Shaped like a sandal or slipper, as in *Dendrobium crepidatum*

crepitans *KREP-ih-tanz*
Rustling; crackling, as in *Hura crepitans*

Hippeastrum striatum (syn. *H. crocatum*)

cretaceus *kret-AY-see-us*
cretacea, cretaceum
Relating to chalk, as in *Dianthus cretaceus*

creticus *KRET-ih-kus*
cretica, creticum
Connected with Crete, Greece, as in *Pteris cretica*

crinitus *krin-EE-tus*
crinita, crinitum
With long, weak hairs, as in *Acanthophoenix crinita*

crispatus *kriss-PAH-tus*
crispata, crispatum
crispus *KRISP-us*
crispa, crispum
Closely curled, as in *Mentha crispa*

cristatus *kris-TAH-tus*
cristata, cristatum
With tassel-like tips, as in *Iris cristata*

crithmifolius *krith-mih-FOH-lee-us*
crithmifolia, crithmifolium
With leaves like *Crithmum*, as in *Achillea crithmifolia*

crocatus *kroh-KAH-tus*
crocata, crocatum
croceus *KRO-kee-us*
crocea, croceum
Saffron yellow, as in *Tritonia crocata*

crocosmiiflorus *kroh-koz-mee-eye-FLOR-us*
crocosmiiflora, crocosmiiflorum
With flowers like *Crocosmia*. The genus of *Crocosmia × crocosmiiflora*
was originally *Montbretia*; the name therefore meant crocosmia-
flowered montbretia

cruciatus *kruks-ee-AH-tus*
cruciata, crusiatum
In the shape of a cross, as in *Gentiana cruciata*

cruentus *kroo-EN-tus*
cruenta, cruentum
Bloody, as in *Lycaste cruenta*

crus-galli *krus GAL-ee*
Cock's spur, as in *Crataegus crus-galli*

crustatus *krus-TAH-tus*
crustata, crustatum
Encrusted, as in *Saxifraga crustata*

LATIN IN ACTION

Also known by the common names sun rose and rock rose, *Helianthemum* will bloom the summer long if sited in an open and sunny position. They are particularly suited to rock gardens, although they can become invasive. As its name suggests, this species has reddish, copper-coloured flowers that change to a darker shade of orange towards the centre. Like many helianthemums, it has sun-tolerant grey-green leaves.

Helianthemum cupreum, rock rose

crystallinus *kris-tal-EE-nus*
crystallina, crystallinum
Crystalline, as in *Anthurium crystallinum*

cucullatus *kuk-yoo-LAH-tus*
cucullata, cucullatum
Like a hood, as in *Viola cucullata*

cucumerifolius *ku-ku-mer-ee-FOH-lee-us*
cucumerifolia, cucumerifolium
With leaves like a cucumber, as in *Cissus cucumerifolia*

cucumerinus *ku-ku-mer-EE-nus*
cucumerina, cucumerinum
Like a cucumber, as in *Trichosanthes cucumerina*

cultorum *kult-OR-um*
Relating to gardens, as in *Trollius × cultorum*

cultratus *kul-TRAH-tus*
cultrata, cultratum
cultriformis *kul-tre-FOR-mis*
cultriformis, cultriforme
Shaped like a knife, as in *Angraecum cultriforme*

cuneatus *kew-nee-AH-tus*
cuneata, cuneatum
In the shape of a wedge, as in *Prostanthera cuneata*

cuneifolius *kew-nee-FOH-lee-us*
cuneifolia, cuneifolium
With leaves shaped like a wedge, as in *Primula cuneifolia*

cuneiformis *kew-nee-FOR-mis*
cuneiformis, cuneiforme
In the form of a wedge, as in *Hibbertia cuneiformis*

cunninghamianus *kun-ing-ham-ee-AH-nus*
cunninghamiana, cunninghamianum
cunninghamii *kun-ing-ham-eye*
May commemorate various people called Cunningham, including Alan Cunningham (1791–1839), English plant collector and botanist, as in *Archontophoenix cunninghamiana*

cupreatus *kew-pree-AH-tus*
cupreata, cupreatum
cupreus *kew-pree-US*
cuprea, cupreum
The colour of copper, as in *Alocasia cuprea*

cupressinus *koo-pres-EE-nus*
cupressina, cupressinum
cupressoides *koo-press-OY-deez*
Resembling cypress, as in *Fitzroya cupressoides*

curassavicus *ku-ra-SAV-ih-kus*
curassavica, curassavicum
From Curaçao, Lesser Antilles, as in *Asclepias curassavica*

curtus *KUR-tus*
curta, curtum
Shortened, as in *Ixia curta*

curvatus *KUR-va-tus*
curvata, curvatum
Curved, as in *Adiantum curvatum*

curvifolius *kur-vi-FOH-lee-us*
curvifolia, curvifolium
With curved leaves, as in *Ascocentrum curvifolium*

cuspidatus *kus-pi-DAH-tus*
cuspidata, cuspidatum
With a stiff point, as in *Taxus cuspidata*

cuspidifolius *kus-pi-di-FOH-lee-us*
cuspidifolia, cuspidifolium
With leaves with a stiff point, as in *Passiflora cuspidifolia*

cyananthus *sy-an-NAN-thus*
cyanantha, cyananthum
With blue flowers, as in *Penstemon cyananthus*

cyaneus *sy-AN-ee-us*
cyanea, cyaneum
cyanus *sy-AH-nus*
Blue, as in *Allium cyaneum*

cyanocarpus *sy-an-o-KAR-pus*
cyanocarpa, cyanocarpum
Bearing blue fruit, as in *Rhododendron cyanocarpum*

cyatheoides *sigh-ath-ee-OY-deez*
Resembling *Cyathea*, as in *Sadleria cyatheoides*

cyclamineus *SIGH-kluh-min-ee-us*
cyclaminea, cyclamineum
Like *Cyclamen*, as in *Narcissus cyclamineus*

cyclocarpus *sigh-klo-KAR-pus*
cyclocarpa, cyclocarpum
With fruit arranged in a circle, as in *Enterolobium cyclocarpum*

cylindraceus *sil-in-DRA-see-us*
cylindracea, cylindraceum
cylindricus *sil-IN-drih-kus*
cylindrica, cylindricum
Long and cylindrical, as in *Vaccinium cylindraceum*

cylindrostachyus *sil-in-dro-STAK-ee-us*
cylindrostachya, cylindrostachyum
With a cylindrical spike, as in *Betula cylindrostachya*

cymbalaria *sim-buh-LAR-ee-uh*
Like *Cymbalaria*, especially its leaves, as in *Ranunculus cymbalaria*

cymbiformis *sim-BIH-for-mis*
cymbiformis, cymbiforme
Shaped like a boat, as in *Haworthia cymbiformis*

cymosus *sy-MOH-sus*
cymosa, cymosum
With flower clusters that flower from the centre outwards,
as in *Rosa cymosa*

cynaroides *sin-nar-OY-deez*
Resembling *Cynara*, as in *Protea cynaroides*

cyparissias *sy-pah-RIS-ee-as*
Latin name for a kind of spurge, as in *Euphorbia cyparissias*

cyprius *SIP-ree-us*
cypria, cyprium
Connected with the island of Cyprus, as in *Cistus × cyprius*

cytisoides *sit-iss-OY-deez*
Resembling broom (*Cytisus*), as in *Lotus cytisoides*

D

dactyliferus *dak-ty-LIH-fer-us*
dactylifera, dactyliferum
With fingers; finger-like, as in *Phoenix dactylifera*

dactyloides *dak-ty-LOY-deez*
Resembling fingers, as in *Hakea dactyloides*

dahuricus *da-HYUR-ih-kus*
dahurica, dahuricum
Connected with Dahuria, a region incorporating parts of Siberia and Mongolia, as in *Codonopsis dahurica*

dalhousiae *dal-HOO-zee-ay*
dalhousieae
Named after Susan Georgiana Ramsay, Marchioness of Dalhousie (1817–1853), as in *Rhododendron dalhousiae*

dalmaticus *dal-MAT-ih-kus*
dalmatica, dalmaticum
Connected with Dalmatia, Croatia, as in *Geranium dalmaticum*

damascenus *dam-ASK-ee-nus*
damascena, damascenum
Connected with Damascus, Syria, as in *Nigella damascena*

dammeri *DAM-mer-ee*
Named after Carl Lebrecht Udo Dammer (1860–1920), German botanist, as in *Cotoneaster dammeri*

danfordiae *dan-FORD-ee-ay*
Named after Mrs C.G. Danford, 19th-century traveller, as in *Iris danfordiae*

danicus *DAN-ih-kus*
danica, danicum
From Denmark, as in *Erodium danicum*

daphnoides *daf-NOY-deez*
Resembling *Daphne*, as in *Salix daphnoides*

darleyensis *dar-lee-EN-sis*
Of Darley Dale nursery (James Smith & Sons), Derbyshire, as in *Erica × darleyensis*

darwinii *dar-WIN-ee-eye*
Named after Charles Darwin (1809–82), English naturalist, as in *Berberis darwinii*

LATIN IN ACTION

Many gardeners claim that the Damask rose has the best fragrance of all the old roses. Traditionally the highly fragranced essential oil attar of roses has been extracted from the flowers of the Damask. This rose was officially introduced into Europe from Persia in the 13th century, but it may have arrived much earlier as it is thought to appear in Roman frescoes. The lovely light pink flowers of 'Bella Donna' are borne in clusters and have an open and lax habit. The plant has grey-green leaves and very bristly and thorny stems. These vigorous and robust hardy shrubs are far easier to grow than many other roses. The group known as Summer Damasks flower only once, while the Autumn Damasks repeat flower later in the year. The blooms are followed by long and slender hips in the autumn.

Rosa 'Bella Donna', Damask rose

dasyacanthus *day-see-uh-KAN-thus*
dasyacantha, dasyacanthum
With thick spines, as in *Escobaria dasyacantha*

dasyanthus *day-see-AN-thus*
dasyantha, dasyanthum
With shaggy flowers, as in *Spiraea dasyantha*

dasycarpus *day-see-KAR-pus*
dasycarpa, dasycarpum
With hairy fruit, as in *Angraecum dasycarpum*

dasyphyllus *das-ee-FIL-us*
dasyphylla, dasyphyllum
With shaggy leaves, as in *Sedum dasyphyllum*

dasystemon *day-see-STEE-mon*
With hairy stamens, as in *Tulipa dasystemon*

daucifolius *daw-ke-FOH-lee-us*
daucifolia, daucifolium
With leaves like carrot (*Daucus*), as in *Asplenium daucifolium*

LATIN IN ACTION

Named by Linnaeus in his *Species Plantarum* of 1753,
this rhododendron is native to Siberia, Mongolia and
northern China and Japan. Very hardy, it is a compact
and deciduous shrub, although some leaves may
persist through winter (hence *sempervirens*, meaning
evergreen). Its name refers to the Dauria region of
southeastern Siberia.

Rhododendron ledebourii
(syn. *R. dauricum* var. *sempervirens*)

---

daucoides *do-KOY-deez*
Resembling carrot (*Daucus*), as in *Erodium daucoides*

dauricus *DOR-ih-kus*
daurica, dauricum
Connected with Siberia, as in *Lilium dauricum*

davidianus *duh-vid-ee-AH-nus*
davidiana, davidianum
davidii *duh-vid-ee-eye*
Named after Père Armand David (1826–1900), French naturalist
and missionary, as in *Buddleja davidii*

davuricus *dav-YUR-ih-kus*
davurica, davuricum
Connected with Siberia, as in *Juniperus davurica*

dawsonianus *daw-son-ee-AH-nus*
dawsoniana, dawsonianum
Named after Jackson T. Dawson (1841–1916), the first
Superintendent of the Arnold Arboretum, Boston,
USA, as in *Malus × dawsoniana*

dealbatus *day-al-BAH-tus*
dealbata, dealbatum
Covered with an opaque white powder, as in *Acacia dealbata*

debilis *deb-IL-is*
debilis, debile
Weak and frail, as in *Asarum debile*

decaisneanus *de-kane-ee-AY-us*
decaisneana, decaisneanum
decaisnei *de-KANE-ee-eye*
Named after Joseph Decaisne (1807–82), French botanist, as in
Aralia decaisneana

decandrus *dek-AN-drus*
decandra, decandrum
With ten stamens, as in *Combretum decandrum*

decapetalus *dek-uh-PET-uh-lus*
decapetala, decapetalum
With ten petals, as in *Caesalpinia decapetala*

deciduus *dee-SID-yu-us*
decidua, deciduum
Deciduous, as in *Larix decidua*

decipiens *de-SIP-ee-enz*
Deceptive; not obvious, as in *Sorbus decipiens*

declinatus *dek-lin-AH-tus*
declinata, declinatum
Bending downwards, as in *Cotoneaster declinatus*

decompositus *de-kom-POZ-ee-tus*
decomposita, decompostitum
Divided several times, as in *Paeonia decomposita*

decoratus *dek-kor-RAH-tus*
decorata, decoratum
decorus *dek-kor-RUS*
decora, decorum
Decorative, as in *Rhododendron decorum*

decumanus *dek-yoo-MAH-nus*
decumana, decumanum
Very large, as in *Phlebodium decumanum*

decumbens *de-KUM-benz*
Trailing with upright tips, as in *Correa decumbens*

decurrens *de-KUR-enz*
Running down the stem, as in *Calocedrus decurrens*

decussatus *de-KUSS-ah-tus*
decussata, decussatum
With leaves that are borne in pairs at right angles to each other, as in *Microbiota decussata*

deflexus *de-FLEKS-us*
deflexa, deflexum
Bending sharply downwards, as in *Enkianthus deflexus*

deformis *de-FOR-mis*
deformis, deforme
Deformed; misshapen, as in *Haemanthus deformis*

degronianum *de-gron-ee-AH-num*
Named after Henri Joseph Degron, director of the French Post Office in Yokohama from 1865–1880, as in *Rhododendron degronianum*

dejectus *dee-JEK-tus*
dejecta, dejectum
Debased, as in *Opuntia dejecta*

delavayi *del-uh-VAY-ee*
Named after Père Jean Marie Delavay (1834–1895), French missionary, explorer and botanist, as in *Magnolia delavayi*

delicatus *del-ih-KAH-tus*
delicata, delicatum
Delicate, as in *Dendrobium × delicatum*

Ceratophyllum demersum, hornwort

deliciosus *de-lis-ee-OH-sus*
deliciosa, deliciosum
Delicious, as in *Monstera deliciosa*

delphiniifolius *del-fin-uh-FOH-lee-us*
delphiniifolia, delphiniifolium
With leaves like *Delphinium*, as in *Aconitum delphiniifolium*

deltoides *del-TOY-deez*
deltoideus *el-TOY-dee-us*
deltoidea, deltoideum
Triangular, as in *Dianthus deltoides*

demersus *DEM-er-sus*
demersa, demersum
Living under water, as in *Ceratophyllum demersum*

deminutus *dee-MIN-yoo-tus*
deminuta, deminutum
Small, diminished, as in *Rebutia deminuta*

demissus *dee-MISS-us*
demissa, demissum
Hanging downwards; weak, as in *Cytisus demissus*

dendroides *den-DROY-deez*
dendroideus *den-DROY-dee-us*
dendroidea, dendroideum
Resembling a tree, as in *Sedum dendroideum*

FRANCIS MASSON
(1741–1805)

CARL PETER THUNBERG
(1743–1828)

Scots-born Francis Masson rose through the ranks from working as a lowly under-gardener at London's Kew Gardens to becoming their very first official plant collector. Under the direction of Sir Joseph Banks (see p. 40), Masson set sail aboard Captain James Cook's ship *Resolution* in 1772, bound for the Cape of Good Hope. Once docked at Cape Town, South Africa, Masson left the ship, which was sailing on to Antarctica, and spent the next three years collecting plants and seeds for Kew.

During Masson's first foray into the interior, he crossed the Cape Flats to Paarl, Stellenbosch, the Holland Mountains and on to the hot springs of Swartberg and Swellenden. Protected by an armed mercenary and travelling aboard a wagon pulled by oxen, he found a terrain rich in unknown and exciting plant species. Once safely back in Cape Town, he dispatched his newly discovered treasures to an appreciative Banks back at Kew. Prior to departing on his next expedition, Masson made the acquaintance of the Swedish botanist Carl Peter Thunberg. The two men decided to explore together, this time setting out on horseback and accompanied by wagons carrying supplies and with four assistants. The country they crossed was often treacherous, but the number and variety of plants they found was hugely rewarding. Among the species they encountered was *Brabejum stellatifolium*, *Kiggelaria africana* and *Metrosideros angustifolia*, along with many mountain-dwelling plants not previously described. Despite having little formal education, Masson published his *Stapeliae novae* in 1796, which he not only wrote but also illustrated.

Many of the plants that grace our gardens and conservatories today we owe to Masson's intrepid travels. From his Cape expeditions he brought back

Named after Queen Charlotte of England, the exotic looking bird-of-paradise, *Strelitzia reginae*, is one of the many South African plants that Masson introduced into Britain. *Kniphofia rooperi*, the red-hot poker, and several pelargoniums are also numbered among his finds.

species of *Amaryllis*, *Erica*, *Oxalis*, *Pelargonium* and *Protea*, as well as succulents and bulbs such as *Gladiolus*. The two-leaved and sweet-smelling *Massonia* is named in his honour. He also travelled to the Azores, Madeira, North Africa, Tenerife and the West Indies. Despite encountering such dangers as escaped convicts on Table Mountain, imprisonment by French expeditionary forces on the island of Grenada, and capture by French pirates in the Atlantic, Masson succeeded in introducing over 1,000 new species into Britain. Following a long and impressive career, the last trip he embarked upon was to North America. Among his many journeys was one along the Ottawa River and Lake Superior with traders from the North West Company. He collected living plants, including some aquatics, and seeds that he dispatched back to England. He died in Montreal, Canada, in 1805.

Carl Peter Thunberg, Masson's companion on the second Cape expedition, studied medicine at the University of Uppsala and was a former student of Linnaeus (see p. 132), with whom he had enjoyed botanising trips. In 1770 he travelled to Paris to further his medical studies. There he was invited to travel to Japan to collect plants for the Dutch collector Johannes Burman, although it was difficult for foreigners to gain entry into Japan at that time. One solution was to become an employee of the Dutch East India Company; prior to his departure, the company arranged for Thunberg to spend time in Cape Town to improve his Dutch. In all he spent three years there and met and collected with Masson. The genus *Thunbergia* is named in his honour.

Huernia campanulata

Illustration of *Huernia campanulata* (syn. *Stapelia campanulata*) from Francis Masson's *Stapeliae novae: or, a collection of several new species of that genus; discovered in the interior parts of Africa.*

Thunberg eventually reached Japan, where he collected many new plant species. Thunberg circumvented the travel restrictions placed on foreigners by unofficially trading his knowledge of European medicine for plants. His Japanese specimens are listed in his 1784 collection *Flora Japonica*; his accounts of 21 new genera and hundreds of new species earned him the appellation 'the Japanese Linnaeus'. A later publication, *Flora Capensis*, compiled with the German botanist Joseph Schultes, describes many of the flowers he saw at the Cape. Returning to Sweden, Thunberg became Professor of Botany at Uppsala, a post he inherited from Linnaeus's son Carl von Linné the Younger (1741–83).

'TRAVELLERS WHO OCCASIONALLY MET HIM IN REMOTE COUNTRIES ... AND MEN OF SCIENCE WHO KNEW HIS UNREMITTING BOTANICAL LABOURS AND COULD ESTIMATE HIS TALENTS, BEAR EQUAL TESTIMONY OF HIS MERITS AND THEIR WRITINGS INCONTESTABLY EVINCE HIS VERY UNCOMMON SUCCESS.'

Obituary of Francis Masson, Montreal Gazette

Fuchsia
denticulata

dendrophilus *den-dro-FIL-us*
dendrophila, dendrophilum
Tree-loving, as in *Tecomanthe dendrophila*

dens-canis *denz KAN-is*
Term for dog's tooth, as in *Erythronium dens-canis*

densatus *den-SA-tus*
densata, densatum
densus *den-SUS*
densa, densum
Compact; dense, as in *Trichodiadema densum*

densiflorus *den-see-FLOR-us*
densiflora, densiflorum
Densely flowered, as in *Verbascum densiflorum*

densifolius *den-see-FOH-lee-us*
densifolia, densifolium
Densely leaved, as in *Gladiolus densifolilus*

dentatus *den-TAH-tus*
dentata, dentatum
With teeth, as in *Ligularia dentata*

denticulatus *den-tik-yoo-LAH-tus*
denticulata, denticulatum
Slightly toothed, as in *Primula denticulata*

denudatus *dee-noo-DAH-tus*
denudata, denudatum
Bare; naked, as in *Magnolia denudata*

deodara *dee-oh-DAR-uh*
From the Indian name for the deodar, as in *Cedrus deodara*

depauperatus *de-por-per-AH-tus*
depauperata, depauperatum
Not properly developed; dwarfed, as in *Carex depauperata*

dependens *de-PEN-denz*
Hanging down, as in *Celastrus dependens*

deppeanus *dep-ee-AH-nus*
deppeana, deppeanum
Named after Ferdinand Deppe (1794–1861), German botanist, as in *Juniperus deppeana*

depressus *de-PRESS-us*
depressa, depressum
Flattened or pressed down, as in *Gentiana depressa*

deserti *DES-er-tee*
Connected with the desert, as in *Agave deserti*

desertorum *de-zert-OR-um*
Of deserts, as in *Alyssum desertorum*

detonsus *de-TON-sus*
detonsa, detonsum
Bare; shorn, as in *Gentianopsis detonsa*

deustus *dee-US-tus*
deusta, deustum
Burned, as in *Tritonia deusta*

diabolicus *dy-oh-BOL-ih-kus*
diabolica, diabolicum
Devilish, as in *Acer diabolicum*

diacanthus *dy-ah-KAN-thus*
diacantha, diacanthum
With two spines, as in *Ribes diacanthum*

diadema *dy-uh-DEE-ma*
Crown; diadem, as in *Begonia diadema*

diandrus *dy-AN-drus*
diandra, diandrum
With two stamens, as in *Bromus diandrus*

dianthiflorus *die-AN-thuh-flor-us*
dianthiflora, dianthiflorum
With flowers like pink (*Dianthus*), as in *Episcia dianthiflora*

diaphanus *dy-AF-a-nus*
diaphana, diaphanum
Transparent, as in *Berberis diaphana*

dichotomus *dy-KAW-toh-mus*
dichotoma, dichotomum
In forked pairs, as in *Iris dichotoma*

dichroanthus *dy-kroh-AN-thus*
dichroantha, dichroanthum
With flowers of two quite different colours, as in
Rhododendron dichroanthum

dichromus *dy-Kroh-mus*
dichroma, dichromum
dichrous *dy-KRUS*
dichroa, dichroum
With two distinct colours, as in *Gladiolus dichrous*

dictyophyllus *dik-tee-oh-FIL-us*
dictyophylla, dictyophyllum
With leaves that have a net pattern, as in *Berberis dictyophylla*

didymus *DID-ih-mus*
didyma, didymum
In pairs, twin, as in *Monarda didyma*

difformis *dif-FOR-mis*
difformis, difforme
With an unusual form, unlike the rest of the genus,
as in *Vinca difformis*

diffusus *dy-FEW-sus*
diffusa, diffusum
With a spreading habit, as in *Cyperus diffusus*

digitalis *dij-ee-TAH-lis*
digitalis, digitale
Like a finger, as in *Penstemon digitalis*

digitatus *dig-ee-TAH-tus*
digitata, digitatum
Like the shape of an open hand, as in *Schefflera digitata*

dilatatus *di-la-TAH-tus*
dilatata, dilatatum
Spread out, as in *Dryopteris dilatata*

dilutus *di-LOO-tus*
diluta, dilutum
Diluted (i.e. pale) as in *Alstroemeria diluta*

dimidiatus *dim-id-ee-AH-tus*
dimidiata, dimidiatum
Divided into two different or unequal parts, as in
Asarum dimidiatum

dimorphus *dy-MOR-fus*
dimorpha, dimorphum
With two different forms of leaf, flower or fruit, as in
Ceropegia dimorpha

dioicus *dy-OY-kus*
dioica, dioicum
With the male reproductive organs on one plant and the
female on another, as in *Arunus dioicus*

dipetalus *dy-PET-uh-lus*
dipetala, dipetalum
With two petals, as in *Begonia dipetala*

LATIN IN ACTION

Corydalis are hardy perennials belonging to the family
Papaveraceae. The synonymous epithet *digitata* means
like the shape of an open hand. As the flowers of
Corydalis are long and tubular, this may originally have
been a reference to the shape of the fern-like leaves.

Corydalis solida
(syn. *C. digitata*),
fingered corydalis

DIGITALIS

Among the many apocryphal accounts of how this plant got its vulpine name is the tale of the fairies who sewed the characteristically finger-shaped flowers of the foxglove into gloves for friendly foxes, thus allowing them to leave no incriminating paw prints on the henhouse door. Another tells us the phrase refers to 'folks' gloves', the folks in question being fairies who wore the flowers as gloves. The prevalence of forest-dwelling fairies in these stories probably stems from the natural woodland habitat favoured by *Digitalis*. The Latin name comes from *digitus*, meaning finger, and refers to the finger-like flowers borne on tall stems. The French call them *gant de Notre Dame*, Our Lady's glove, whereas in a sartorial departure, the Irish refer to the plant as fairy's cap.

Belonging to the family *Plantaginaceae*, Digitalis can grow up to 1.5 m (5 ft). In the garden they are best treated as biennials, especially if colour is important, as they tend to deteriorate over time and revert to the common purple strains. Rather misleadingly, *D. purpurea* can be white, cream or pink as well as purple. For a true white, choose *D. purpurea* f. *albiflora* 'Camelot White' or 'Dalmatian White', which look lovely in the twilight. (*Purpureus*, *purpurea*, *purpureum*, purple, and *albiflorus*, *albiflora*, *albiflorum*, with white

Digitalis purpurea,
common foxglove

flowers.) *D. purpurea* has been used as a treatment for heart problems for centuries. However, it can prove toxic if consumed in the wrong quantities; hence another common name – dead man's bells.

Several plant species in other genera are also associated with fingers. *Digitatus*, *digitata* and *digitatum* all relate to five fingers and literally mean in the shape of an open hand. The African tree *Adansonia digitata* has five leaflets that resemble the splayed fingers of a hand. The fan clubmoss, *Diphasiastrum digitatum*, has fan- or palm-like lateral stems, while *Cyperus digitatus* is commonly known as finger flatsedge.

Digitalis lutea,
yellow foxglove

diphyllus *dy-FIL-us*
diphylla, diphyllum
With two leaves, as in *Bulbine diphylla*

dipsaceus *dip-SAK-ee-us*
dipsacea, dipsaceum
Like teasel (*Dipsacus*), as in *Carex dipsacea*

dipterocarpus *dip-ter-oh-KAR-pus*
dipterocarpa, dipterocarpum
With two-winged fruit, as in *Thalictrum dipterocarpum*

dipterus *DIP-ter-us*
diptera, dipterum
With two wings, as in *Halesia diptera*

dipyrenus *dy-pie-REE-nus*
dipyrena, dipyrenum
With two seeds or kernels, as in *Ilex dipyrena*

dis-
Used in compound words to denote apart

disciformis *disk-ee-FOR-mis*
disciformis, disciforme
Shaped like a disc, as in *Medicago disciformis*

discoideus *dis-KOY-dee-us*
discoidea, discoideum
Without rays, as in *Matricaria discoidea*

discolor *DIS-kol-or*
Of two quite different colours, as in *Salvia discolor*

dispar *DIS-par*
Unequal; unusual for a genus, as in *Restio dispar*

dispersus *dis-PER-sus*
dispersa, dispersum
Scattered, as in *Paranomus dispersus*

dissectus *dy-SEK-tus*
dissecta, dissectum
Deeply cut or divided, as in *Cirsium dissectum*

dissimilis *dis-SIM-il-is*
dissimilis, dissimile
Differing from the norm for a particular genus,
as in *Columnea dissimilis*

distachyus *dy-STAK-yus*
distachya, distachyum
With two spikes, as in *Billbergia distachya*

distans *DIS-tanz*
Widely apart, as in *Watsonia distans*

distichophyllus *dis-ti-koh-FIL-us*
distichophylla, distichophyllum
With leaves appearing in two ranks or levels, as in
Buckleya distichophylla

distichus *DIS-tih-kus*
disticha, distichum
In two ranks or levels, as in *Taxodium distichum*

distortus *DIS-tor-tus*
distorta, distortum
Misshapen, as in *Adonis distorta*

distylus *DIS-sty-lus*
distyla, distylum
With two styles, as in *Acer distylum*

diurnus *dy-YUR-nus*
diurna, diurnum
Flowering by day, as in *Cestrum diurnum*

divaricatus *dy-vair-ih-KAH-tus*
divaricata, divaricatum
With a spreading and straggling habit, as in *Phlox divaricata*

Ranunculus asiaticus
var. *discolor*,
Persian buttercup

divergens *div-VER-jenz*
Spreading out a long way from the centre, as in *Ceanothus divergens*

diversifolius *dy-ver-sih-FOH-lee-us*
diversifolia, diversifolium
With diverse leaves, as in *Hibiscus diversifolius*

diversiformis *dy-ver-sih-FOR-mis*
diversiformis, diversiforme
With diverse forms, as in *Romulea diversiformis*

divisus *div-EE-sus*
divisa, divisum
Divided, as in *Pennisetum divisum*

dodecandrus *doh-DEK-an-drus*
dodecandra, dodecandrum
With twelve stamens, as in *Cordia dodecandra*

doerfleri *DOOR-fleur-eye*
Named after Ignaz Dörfler (1866–1950), German botanist,
as in *Colchicum doerfleri*

dolabratus *dol-uh-BRAH-tus*
dolabrata, dolabratum
dolabriformis *doh-la-brih-FOR-mis*
dolabriformis, dolabriforme
Shaped like a hatchet, as in *Thujopsis dolabrata*

dolosus *do-LOH-sus*
dolosa, dolosum
Deceitful; looking like another plant, as in *Cattleya × dolosa*

dombeyi *DOM-bee-eye*
Named after Joseph Dombey (1742–94), French botanist,
as in *Nothofagus dombeyi*

domesticus *doh-MESS-tih-kus*
domestica, domesticum
Domesticated, as in *Malus domestica*

douglasianus *dug-lus-ee-AH-nus*
douglasiana, douglasianum
douglasii *dug-lus-ee-eye*
Named after David Douglas (1799–1834), Scottish plant hunter,
as in *Limnanthes douglasii*

drabifolius *dra-by-FOH-lee-us*
drabifolia, drabifolium
With leaves like whitlow-grass (*Draba*), as in *Centaurea drabifolia*

draco *DRAY-koh*
Dragon, as in *Dracaena draco*

dracunculus *dra-KUN-kyoo-lus*
Small dragon, as in *Artemisia dracunculus*

drummondianus *drum-mond-ee-AH-nus*
drummondiana, drummondianum
drummondii *drum-mond-EE-eye*
Named after either James Drummond (1786–1863) or Thomas
Drummond (1793–1835), brothers who collected plants in Australia
and North America respectively, as in *Phlox drummondii*

drupaceus *droo-PAY-see-us*
drupacea, drupaceum
drupiferus *droo-PIH-fer-us*
drupifera, drupiferum
With fleshy fruits such as peach or cherry, as in *Hakea drupacea*

drynarioides *dri-nar-ee-OY-deez*
Resembling oakleaf fern (*Drynaria*), as in *Aglaomorpha drynarioides*

dubius *DOO-bee-us*
dubia, dubium
Doubtful, unlike the rest of the genus, as in *Ornithogalum dubium*

dulcis *DUL-sis*
dulcis, dulce
Sweet, as in *Prunus dulcis*

dumetorum *doo-met-OR-um*
From hedges or bushes, as in *Fallopia dumetorum*

dumosus *doo-MOH-sus*
dumosa, dumosum
Bushy; shrubby, as in *Alluaudia dumosa*

duplicatus *doo-plih-KAH-tus*
duplicata, duplicatum
Double; duplicate, as in *Brachystelma duplicatum*

durus *DUR-us*
dura, durum
Hard, as in *Blechnum durum*

dyeri *DY-er-eye*
dyerianus *dy-er-ee-AH-nus*
dyeriana, dyerianum
Named after Sir William Turner Thiselton-Dyer (1843–1928),
English botanist and Director of Kew Gardens, London, England,
as in *Strobilanthes dyeriana*

E

e-, ex-
Used in compound words to denote without, out of

ebeneus *eb-en-NAY-us*
ebenea, ebeneum
ebenus *eb-en-US*
ebena, ebenum
Ebony-black, as in *Carex ebenea*

ebracteatus *e-brak-tee-AH-tus*
ebracteata, ebracteatum
Without bracts, as in *Eryngium ebracteatum*

eburneus *eb-URN-ee-us*
eburnea, eburneum
Ivory-white, as in *Angraecum eburneum*

echinatus *ek-in-AH-tus*
echinata, echinatum
With prickles like a hedgehog, as in *Pelargonium echinatum*

echinosepalus *ek-in-oh-SEP-uh-lus*
echinosepala, echinosepalum
With prickly sepals, as in *Begonia echinosepala*

echioides *ek-ee-OY-deez*
Resembling viper's bugloss (*Echium*), as in *Picris echioides*

ecornutus *ek-kor-NOO-tus*
ecornuta, ecornutum
Without horns, as in *Stanhopea ecornuta*

edgeworthianus *edj-wor-thee-AH-nus*
edgeworthiana, edgeworthianum
edgeworthii *edj-WOR-thee-eye*
Named after Michael Pakenham Edgeworth (1812–81) of the East
India Company, as in *Rhododendron edgeworthii*

edulis *ED-yew-lis*
edulis, edule
Edible, as in *Dioon edule*

effusus *eff-YOO-sus*
effusa, effusum
Spreading loosely, as in *Juncus effusus*

elaeagnifolius *el-ee-ag-ne-FOH-lee-us*
elaeagnifolia, elaeagnifolium
With leaves like *Elaeagnus*, as in *Brachyglottis elaeagnifolia*

elasticus *ee-LASS-tih-kus*
elastica, elasticum
Elastic, producing latex, as in *Ficus elastica*

elatus *el-AH-tus*
elata, elatum
Tall, as in *Aralia elata*

elegans *el-ee-GANS*
elegantulus *el-eh-GAN-tyoo-lus*
elegantula, elegantulum
Elegant, as in *Desmodium elegans*

elegantissimus *el-ee-gan-TISS-ih-mus*
elegantissima, elegantissimum
Very elegant, as in *Schefflera elegantissima*

LATIN IN ACTION

This hardy perennial plant certainly lives up to its
name *elatum*, tall, as its long and graceful flower
spikes can reach a stately 2 m (6 ft). Having a stiff and
erect habit it bears large flowers that are often double
or semi-double.

Delphinium elatum,
larkspur

elephantipes *ell-uh-fan-TY-peez*
Resembling an elephant's foot, as in *Yucca elephantipes*

elliottianus *el-ee-ot-ee-AH-nus*
elliottiana, elliottianum
Named after Captain George Henry Elliott (1813–1892), as in
Zantedeschia elliottiana

elliottii *el-ee-ot-EE-eye*
Named after Stephen Elliott (1771–1830), American botanist,
as in *Eragrostis elliottii*

ellipsoidalis *e-lip-soy-DAH-lis*
ellipsoidalis, ellipsoidale
Elliptic, as in *Quercus ellipsoidalis*

ellipticus *ee-LIP-tih-kus*
elliptica, ellipticum
Shaped like an ellipse, as in *Garrya elliptica*

elongatus *ee-long-GAH-tus*
elongata, elongatum
Lengthened; elongated, as in *Mammillaria elongata*

elwesii *el-WEZ-ee-eye*
Named after Henry John Elwes (1846–1922), British plant collector,
one of the inaugural recipients of the Victoria Medal of the Royal
Horticultural Society, as in *Galanthus elwesii*

emarginatus *e-mar-jin-NAH-tus*
emarginata, emarginatum
Slightly notched at the margins, as in *Pinguicula emarginata*

eminens *EM-in-enz*
Eminent; prominent, as in *Sorbus eminens*

empetrifolius *em-pet-rih-FOH-lee-us*
empetrifolia, empetrifolium
With leaves like crowberry (*Empetrum*), as in *Berberis empetrifolia*

encliandrus *en-klee-AN-drus*
encliandra, encliandrum
With half the stamens inside the flower tube, as in *Fuchsia encliandra*

endresii *en-DRESS-ee-eye*
endressii
Named after Philip Anton Christoph Endress (1806–31),
German plant collector, as in *Geranium endressii*

engelmannii *en-gel-MAH-nee-eye*
Named after Georg Engelmann (1809–84), German-born
physician and botanist, as in *Picea engelmannii*

Eugenia elliptica

enneacanthus *en-nee-uh-KAN-thus*
enneacantha, enneacanthum
With nine spines, as in *Echinocereus enneacanthus*

enneaphyllus *en-nee-a-FIL-us*
enneaphylla, ennephyllum
With nine leaves or leaflets, as in *Oxalis enneaphylla*

ensatus *en-SA-tus*
ensata, ensatum
In the shape of a sword, as in *Iris ensata*

ensifolius *en-see-FOH-lee-us*
ensifolia, ensifolium
With leaves shaped like a sword, as in *Kniphofia ensifolia*

ensiformis *en-see-FOR-mis*
ensiformis, ensiforme
In the shape of a sword, as in *Pteris ensiformis*

epipactis *ep-ih-PAK-tis*
Greek name for a plant thought to curdle milk, as in *Hacquetia epipactis*

epiphyllus *ep-ih-FIL-us*
epiphylla, epiphyllum
On the leaf (e.g. flowers), as in *Saxifraga epiphylla*

epiphyticus *ep-ih-FIT-ih-kus*
epiphytica, epiphyticum
Growing on another plant, as in *Cyrtanthus epiphyticus*

equestris *e-KWES-tris*
equestris, equestre
equinus *e-KWEE-nus*
equina, equinum
Relating to horses, equestrian, as in *Phalaenopsis equestris*

equisetifolius *ek-wih-set-ih-FOH-lee-us*
equisetifolia, equisetifolium
equisetiformis *eck-kwiss-ee-tih-FOR-mis*
equisetiformis, equisetiforme
Resembling horsetail (*Equisetum*), as in *Russelia equisetiformis*

erectus *ee-RECK-tus*
erecta, erectum
Erect; upright, as in *Trillium erectum*

eri-
Used in compound words to denote woolly

eriantherus *er-ee-AN-ther-uz*
erianthera, eriantherum
With woolly anthers, as in *Penstemon eriantherus*

erianthus *er-ee-AN-thus*
eriantha, erianthum
With woolly flowers, as in *Kohleria eriantha*

ericifolius *er-ik-ih-FOH-lee-us*
ericifolia, ericifolium
With leaves like heather (*Erica*), as in *Banksia ericifolia*

ericoides *er-ik-OY-deez*
Resembling heather (*Erica*), as in *Aster ericoides*

erinaceus *er-in-uh-SEE-us*
erinacea, erinaceum
Like a hedgehog, as in *Dianthus erinaceus*

erinus *er-EE-nus*
Greek name for a plant probably basil, as in *Lobelia erinus*

eriocarpus *er-ee-oh-KAR-pus*
eriocarpa, eriocarpum
With woolly fruit, as in *Pittosporum eriocarpum*

eriocephalus *er-ri-oh-SEF-uh-lus*
eriocephala, eriocephalum
With a woolly head, as in *Lamium eriocephalum*

eriostemon *er-ree-oh-STEE-mon*
With woolly stamens, as in *Geranium eriostemon*

erosus *e-ROH-sus*
erosa, erosum
Jagged, as in *Cissus erosa*

erubescens *er-oo-BESS-enz*
Becoming red; blushing, as in *Philodendron erubescens*

erythro-
Used in compound words to denote red

erythrocarpus *er-ee-throw-KAR-pus*
erythrocarpa, erythrocarpum
With red fruit, as in *Actinidia erythrocarpa*

erythropodus *er-ee-THROW-pod-us*
erythropoda, erythropodum
With a red stem, as in *Alchemilla erythropoda*

erythrosorus *er-rith-roh-SOR-us*
erythrosora, erythrosorum
Having red spore cases, as in *Dryopteris erythrosora*

esculentus *es-kew-LEN-tus*
esculenta, esculentum
Edible, as in *Colocasia esculenta*

etruscus *ee-TRUSS-kus*
estrusca, estruscum
Connected with Tuscany, Italy, as in *Crocus etruscus*

eucalyptifolius *yoo-kuh-lip-tih-FOH-lee-us*
eucalyptifolia, eucalyptifolium
With leaves like *Eucalyptus*, as in *Leucadendron eucalyptifolium*

euchlorus *YOO-klor-us*
euchlora, euchlorum
A healthy green, as in *Tilia × euchlora*

eugenioides *yoo-jee-nee-OY-deez*
Resembling the genus *Eugenia*, as in *Pittosporum eugenioides*

ERYNGIUM

Unsurprisingly perhaps, given its common name, sea holly, and its Latin name, *Eryngium maritimum* originates from coastal regions and grows very happily in well-drained sandy soils in an open sunny site. *Maritimus* (*maritima, maritimum*) means of or relating to the sea. Reaching 30 cm (1 ft) high and bearing metallic-blue flowers that rise gracefully from silver-grey foliage, *E. maritimum* is the perfect plant for coastal gardens. The towering *E. gigantea* (*giganteus, gigantea, giganteum*, unusually large or tall) grows to a stately 1–1.2 m (3–4 ft) high and was so beloved by the English gardener Ellen Willmott (1858–1934) that she would fill her pockets with its seeds then scatter them freely among the flower borders of unsuspecting gardening friends – hence the plant is known today by the common name Miss Willmott's ghost.

Alongside her activities as an early guerrilla gardener, Willmott created extensive gardens at her home Warley Place in Essex and sponsored the expeditions of plant hunters such as Ernest Henry Wilson. It is suggested that at one time she employed as many as 100 gardeners. Indeed, her garden-making mania (which also extended to properties in France and Italy) resulted in the eventual loss of her family fortune and she died in great debt.

Several plants bear her name, including *Potentilla nepalensis* 'Miss Willmott' and *Lilium davidii* var. *willmottiae*.

***Eryngium maritimum*,**
sea holly

Prior to the 18th century, the sweet and aromatic roots of sea holly, *Eryngium maritimum*, were candied and eaten. As well as being tasty they were thought to have aphrodisiac properties.

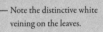

— Note the distinctive white veining on the leaves.

Belonging to the family *Apiaceae*, eryngiums are herbaceous biennials and perennials. Most of them are hardy and have spiny foliage with flower heads surrounded by spine-toothed bracts. *E. giganteum* produces a mat of fleshy green leaves in its first year; these become much stiffer and more prickly in the second year when the flower-bearing stems emerge. The plant then dies after flowering. Cut the flower stems before they begin to fade as, once dried, they make attractive and long-lasting indoor decorations.

eupatorioides *yoo-puh-TOR-ee-oy-deez*
Resembling *Eupatorium*, as in *Agrimonia eupatoria*

euphorbioides *yoo-for-bee-OY-deez*
Resembling spurge (*Euphorbia*), as in *Neobuxbaumia euphorbioides*

europaeus *yoo-ROH-pay-us*
europaea, europaeum
Connected with Europe, as in *Euonymus europaeus*

evansianus *eh-vanz-ee-AH-nus*
evansiana, evansianum
evansii *eh-VANS-ee-eye*
Named after various people called Evans, including Thomas Evans (1751–1814), as in *Begonia grandis* subsp. *evansiana*

exaltatus *eks-all-TAH-tus*
exaltata, exaltatum
Very tall, as in *Nephrolepis exaltata*

exaratus *ex-a-RAH-tus*
exarata, exaratum
Engraved; furrowed, as in *Agrostis exarata*

excavatus *ek-ska-VAH-tus*
excavata, excavatum
Hollowed out, as in *Calochortus excavatus*

excellens *ek-SEL-lenz*
Excellent, as in *Sarracenia* × *excellens*

excelsior *eks-SEL-see-or*
Taller, as in *Fraxinus excelsior*

excelsus *ek-SEL-sus*
excelsa, excelsum
Tall, as in *Araucaria excelsa*

excisus *eks-SIZE-us*
excisa, excisum
Cut away; cut out, as in *Adiantum excisum*

excorticatus *eks-kor-tih-KAH-tus*
excorticata, excorticatum
Lacking or stripped of bark, as in *Fuchsia excorticata*

exiguus *eks-IG-yoo-us*
exigua, exiguum
Very little; poor, as in *Salix exigua*

eximius *eks-IM-mee-us*
eximia, eximium
Distinguished, as in *Eucalyptus eximia*

exoniensis *eks-oh-nee-EN-sis*
exoniensis, exoniense
From Exeter, England, as in *Passiflora* × *exoniensis*

expansus *ek-SPAN-sus*
expansa, expansum
Expanded, as in *Catasetum expansum*

exsertus *ek-SER-tus*
exserta, exsertum
Protruding, as in *Acianthus exsertus*

extensus *eks-TEN-sus*
extensa, extensum
Extended, as in *Acacia extensa*

eyriesii *eye-REE-see-eye*
Named after Alexander Eyries, French 19th-century cactus collector, as in *Echinopsis eyriesii*

Euonymus europaeus,
spindle

EUCALYPTUS

A member of the *Myrtaceae* family, the genus *Eucalyptus* gets its name from the Greek *eu*, meaning well and *kalypto*, meaning to cover. This alludes to the calyx that covers the plant's very distinctive flowers and looks like a lid or a hat. *Eucalyptus* are mostly native to Australia and Tasmania, but are now widely grown worldwide. There are many hundreds of species and they include some of the tallest-growing of all trees. Common names include blue tree, gum tree and string bark tree. *Eucalyptus globulus* is the blue gum or Tasmanian gum. It was first collected on the southeastern coast of Tasmania in the early 1790s by French botanist Jacques-Julien Houtou de Labillardière (1755–1834) and is now the floral emblem of the island.

Eucalyptus are very ornamental, grown for their flowers and aromatic

Eucalyptus pulverulenta, whose specific epithet means 'appearing to be covered in dust'.

leaves. Many species also have attractive bark, most notably *E. dalrympleana*. Its young bark is white or cream-coloured then changes to pink or light brown as it matures. The peeling away of the bark in long strips is characteristic of the genus. *E. camaldulensis*, known as the Murray red gum or the river red gum, is drought-resistant and, like most species of *Eucalyptus*, very fast-growing. *Muellerina eucalyptoides*, also known as creeping mistletoe, has eucalyptus-like leaves and lives as a parasite on *E. camaldulensis* (*eucalyptoides* means resembling *Eucalyptus*). Those who live in warm climates can grow *E. citriodora* for fragrance (*citriodorus, citriodora, citriodorum*, with the scent of lemons). *E. camphora* (*camphorus, camphora, camphorum*, with the smell of camphor) has reddish leaves that are rich in oils smelling faintly of camphor; the tree is often referred to as broad-leaf Sally or the mountain swamp gum.

Eucalyptus require full sun and a well-drained soil and, due to their original native habitat, many tolerate drought well but do badly in wet and cold soils. Avoid exposed, windy spots as these tall trees can easily succumb to windthrow. They do not transplant well, so the careful selection of a suitable site is imperative if the plant is to thrive.

The hat-like calyx from which the Latin name *Eucalyptus* comes.

Eucalyptus globulus, Tasmanian gum.

F

fabaceus *fab-AY-see-us*
fabacea, fabaceum
Like a broad bean, as in *Marah fabacea*

facetus *fa-CEE-tus*
faceta, facetum
Elegant, as *Rhododendron facetum*

fagifolius *fag-ih-FOH-lee-us*
fagifolia, fagifolium
With leaves like beech (*Fagus*), as in *Clethra fagifolia*

falcatus *fal-KAH-tus*
falcata, falcatum
Shaped like a sickle, as in *Cyrtanthus falcatus*

falcifolius *fal-sih-FOH-lee-us*
falcifolia, falcifolium
With leaves in the shape of a sickle, as in *Allium falcifolium*

falciformis *fal-sif-FOR-mis*
falciformis, falciforme
Shaped like a sickle, as in *Falcatifolium falciforme*

falcinellus *fal-sin-NELL-us*
falcinella, falcinellum
Like a small sickle, as in *Polystichum falcinellum*

fallax *FAL-laks*
Deceptive; false, as in *Crassula fallax*

farinaceus *far-ih-NAH-kee-us*
farinacea, farinaceum
Producing starch; mealy, like flour, as in *Salvia farinacea*

farinosus *far-ih-NOH-sus*
farinosa, farinosum
Mealy; powdery, as in *Rhododendron farinosum*

farnesianus *far-nee-zee-AH-nus*
farnesiana, farnesianum
Connected with the Farnese Gardens, Rome, Italy, as in *Acacia farnesiana*

farreri *far-REY-ree*
Named after Reginald Farrer (1880–1920), English plant hunter and botanist, as in *Viburnum farreri*

fasciatus *fash-ee-AH-tus*
fasciata, fasciatum
Bound together, as in *Aechmea fasciata*

fascicularis *fas-sik-yoo-LAH-ris*
fascicularis, fasciculare
fasciculatus *fas-sik-yoo-LAH-tus*
fasciculata, fasciculatum
Clustered or grouped together in bundles, as in *Ribes fasciculatum*

fastigiatus *fas-tij-ee-AH-tus*
fastigiata, fastigiatum
With erect, upright branches, often creating the effect of a column, as in *Cotoneaster fastigiatus*

fastuosus *fast-yoo-OH-sus*
fastuosa, fastuosum
Proud, as in *Cassia fastuosa*

fatuus *FAT-yoo-us*
fatua, fatuum
Insipid; poor quality, as in *Avena fatua*

febrifugus *feb-ri-FEW-gus*
febrifuga, febrifugum
Can reduce fever, as in *Dichroa febrifuga*

fecundus *feh-KUN-dus*
fecunda, fecundum
Fertile; fruitful, as in *Aeschynanthus fecundus*

fejeensis *fee-jee-EN-sis*
fejeensis, fejeense
From the Fiji Islands, South Pacific, as in *Davallia fejeensis*

Gentiana farreri

The Colour of Plants

A sizeable proportion of Latin names belong to the group of terms that are used to describe the colour of plants. The ancient Romans produced their dyes from natural sources such as plants, animals, marine creatures and insects. For those colours that they could quite readily create, such as varying shades of red and yellow, they had plentiful and often rather expansive names. By contrast, blue and green dyes were more difficult to produce; there were also fewer words for greys and browns. As a result, the classical vocabulary is too small for modern scientific needs, and botanists have introduced many new colour terms, including some from Germanic roots, such as *brunneus*, brown.

The accuracy of a colour description that was originally applied to a plant long ago cannot always be relied upon and there will always be an element of subjectivity when describing something as subtle as a colour. What one botanist had come to regard as rose pink may have been a very similar colour to what another would identify as flesh pink. Often the colour name given to a particular plant may in reality be little more than an approximation.

The colour yellow commands a large number of Latin appellations; to make his or her life easier, the gardener should remember that words beginning with *flav-* mostly indicate yellow of some kind. For instance, *Aquilegia flavescens* is the yellow columbine. Trying to remember colour terms in greater detail can be bewildering: for example, *flavens, flaveolus* (*flaveola, flaveolum*), *flavescens, flavidus* (*flavida, flavidum*) all mean yellowish, while *flavus* (*flava, flavum*) is pure yellow. Moving on to other letters of the alphabet, *luteus* (*lutea, luteum*) also means yellow; *luridus* (*lurida, luridum*) means pale yellow, and *luteolus* (*luteola, luteolum*) is yet another term for yellowish. To further complicate matters, the prefix *xanth-* also refers to yellow-coloured things, including plants with yellow fruits, *xanthocarpus* (*xanthocarpa, xanthocarpum*), yellow nerves, *xanthonervis* (*xanthonervis, xanthonerve*), and yellow roots, *xanthorrhizus* (*xanthorrhiza, xanthorrhizum*).

There are several colour terms relating to green. *Viri-* at the beginning of a word often denotes green, such as *virens*, green; *virescens*, light green; *viridescens*, greenish, and so on. *Narcissus viridiflorus* has unusual flowers that have a distinctive green tinge (*viridiflorus, viridiflora, viridiflorum*, with green flowers). *Viridifuscus* (*viridifusca, viridifuscum*) means

Pelargonium bicolor,
two-colour storksbill

As the common name indicates, the two-colour storksbill is an example of a plant with bicolor flowers.

Ixia viridiflora,
turquoise ixia

Green-flowered plants are perhaps less obvious choices
for garden use; this is one of the most beautiful.

Nymphaea alba,
white water lily

Alba denotes white, as with the pure white
flowers of this water lily.

green-brown. There is also a range of terms for
blue-green. *Glaucescens* means with a bloom or blue/
green colour. If leaves appear to be bluish green or
have a bloom they are called *glaucophyllus* (*glaucophylla, glaucophyllum*), as in *Clematis glaucophylla*.

Considerable subtlety is displayed when
describing some colours, most noticeably white and
silver. Along with the various terms for white such as
albidus (*albida, albidum*) and *albus* (*alba, album*),
there are those that describe a plant as white as snow
or as growing near snow. These are *nivalis* (*nivalis,
nivale*), *niveus* (*nivea, niveum*) and *nivosus* (*nivosa,
nivosum*). Both prefixes *argent-* and *argyro-* mean
silver. Thus one finds *argenteoguttatus* (*argenteoguttata, argenteoguttatum*) to describe a plant with silver
spots and *argyroneurus* (*argyroneura, argyroneurum*)

one with silver veins. One also finds these epithets appearing as cultivar names, as with the cultivar *Begonia
arborescens* var. *arborescens* 'Argenteoguttata', an
attractive plant with distinctive white-spotted leaves.

More general terms that relate to colouring
include *bicolor*, meaning a plant that has two colours
– such as *Rosa foetida* 'Bicolor', which has scarlet
flowers with a yellow underside. Likewise, *discolor*
indicates a part of a plant has two colours – for
instance *Cissus discolor*, the rex begonia vine with its
green- and silver-coloured leaves that are red
underneath. In contrast, *concolor* means all the same
colour, like *Abies concolor*, the white fir. *Tinctorius*
(*tinctoria, tinctorium*) relates to dyes, as in the dried
flowers of safflower, *Carthamus tinctorius*, which are
used to produce natural yellow dyes.

fennicus *FEN-nih-kus*
fennica, fennicum
Connected with Finland, as in *Picea fennica*

fenestralis *fen-ESS-tra-lis*
fenestralis, fenestrale
With openings like a window, as in *Vriesea fenestralis*

ferax *FER-aks*
Fruitful, as in *Fargesia ferax*

ferox *FER-oks*
Ferocious; thorny, as in *Datura ferox*

ferreus *FER-ee-us*
ferrea, ferreum
Connected with iron; hard as iron, as in *Caesalpinia ferrea*

ferrugineus *fer-oo-GIN-ee-us*
ferruginea, ferrugineum
Rust-coloured, as in *Digitalis ferruginea*

fertilis *fer-TIL-is*
fertilis, fertile
With plenty of fruit; with many seeds, as in *Robinia fertilis*

festalis *FES-tuh-lis*
festalis, festale
festivus *fes-TEE-vus*
festiva, festivum
Festive; bright, as in *Hymenocallis × festalis*

fibrillosus *fy-BRIL-oh-sus*
fibrillosa, fibrillosum
fibrosus *fy-BROH-sus*
fibrosa, fibrosum
Fibrous, as in *Dicksonia fibrosa*

ficifolius *fik-ee-FOH-lee-us*
ficifolia, ficifolium
With fig-like leaves, as in *Cucumis ficifolius*

ficoides *fy-KOY-deez*
ficoideus *fy-KOY-dee-us*
ficoidea, ficoideum
Resembling a fig (*Ficus*), as in *Senecio ficoides*

filamentosus *fil-uh-men-TOH-sus*
filamentosa, filamentosum
filarius *fil-AH-ree-us*
filaria, filarium
With filaments or threads, as in *Yucca filamentosa*

filicinus *fil-ih-SEE-nus*
filicina, filicinum
filiculoides *fil-ih-kyu-LOY-deez*
Resembling a fern, as in *Asparagus filicinus*

filicifolius *fil-ee-kee-FOH-lee-us*
filicifolia, filicifolium
With leaves like a fern, as in *Polyscias filicifolia*

fili-
Used in compound words to denote thread-like

filicaulis *fil-ee-KAW-lis*
filicaulis, filicaule
With a thread-like stem, as in *Alchemilla filicaulis*

filipes *fil-EE-pays*
With thread-like stalks, as in *Rosa filipes*

filipendulus *fil-ih-PEN-dyoo-lus*
filipendula, filipendulum
Like meadow-sweet (*Filipendula*), as in *Oenanthe filipendula*

fimbriatus *fim-bry-AH-tus*
fimbriata, fimbriatum
Fringed, as in *Silene fimbriata*

firmatus *fir-MAH-tus*
firmata, firmatum
firmus *fir-MUS*
firma, firmum
Strong, as in *Abies firma*

fissilis *FISS-ill-is*
fissilis, fissile
fissus *FISS-us*
fissa, fissum
fissuratus *fis-zhur-RAH-tus*
fissurata, fissuratum
With a split, as in *Alchemilla fissa*

fistulosus *fist-yoo-LOH-sus*
fistulosa, fistulosum
Hollow, as in *Asphodelus fistulosus*

flabellatus *fla-bel-AH-tus*
flabellata, flabellatum
Like an open fan, as in *Aquilegia flabellata*

flabellifer *fla-BEL-lif-er*
flabellifera, flabelliferum
Bearing a fan-like structure, as in *Borassus flabellifer*

flabelliformis *fla-bel-ih-FOR-mis*
flabelliformis, flabelliforme
Shaped like a fan, as in *Erythrina flabelliformis*

flaccidus *FLA-sih-dus*
flaccida, flaccidum
Weak: soft; feeble, as in *Yucca flaccida*

flagellaris *fla-gel-AH-ris*
flagellaris, flagellare
flagelliformis *fla-gel-ih-FOR-mis*
flagelliformis, flagelliforme
Like a whip; with long, thin shoots, as in *Celastrus flagellaris*

flammeus *FLAM-ee-us*
flammea, flammeum
Flame-coloured; flame-like, as in *Tigridia flammea*

flavens *flav-ENZ*
flaveolus *fla-VEE-oh-lus*
flaveola, flaveolum
flavescens *flav-ES-enz*
flavidus *FLA-vid-us*
flavida, flavidum
Various kinds of yellow (see p. 86), as in *Anigozanthos flavidus*

flavicomus *flay-vih-KOH-mus*
flavicoma, flavicomum
With yellow hair, as in *Euphorbia flavicoma*

flavissimus *flav-ISS-ih-mus*
flavissima, flavissimum
Deep yellow, as in *Zephyranthes flavissima*

flavovirens *fla-voh-VY-renz*
Greenish yellow, as in *Callistemon flavovirens*

flavus *FLA-vus*
flava, flavum
Pure yellow, as in *Crocus flavus*

flexicaulis *fleks-ih-KAW-lis*
flexicaulis, flexicaule
With a supple stem, as in *Strobilanthes flexicaulis*

flexilis *FLEKS-il-is*
flexilis, flexile
Pliant, as in *Pinus flexilis*

flexuosus *fleks-yoo-OH-sus*
flexuosa, flexuosum
Indirect; zigzagging, as in *Corydalis flexuosa*

LATIN IN ACTION

The epithet *fistulosum* means hollow; in this instance it relates to the stems of the plant. The genus name *Oenanthe* derives from Greek and means wine flower; when the stems are crushed, they release a smell that resembles wine. The flowers appear in tight umbels and from a distance look rather like those of scabious. Evergreen and hardy *Oenanthe fistulosa* is a native wildflower of the British Isles; its natural habitat is marshy ground, shallow water or boggy sites. Its grey-green leaves resemble those of carrot and it grows to a height of 60 cm (2 ft). *O. fistulosa* makes an attractive pond-side plant. Do beware, as all the *Oenanthes* are poisonous. Indeed, *O. crocata*, commonly known as the hemlock water dropwort, is possibly the most poisonous British plant of all.

Oenanthe fistulosa,
tubular dropwort

floccigerus *flok-KEE-jer-us*
floccigera, floccigerum
floccosus *flok-KOH-sus*
floccosa, floccosum
With a woolly texture, as in *Rhipsalis floccosa*

florentinus *flor-en-TEE-nus*
florentina, florentinum
Connected with Florence, Italy, as in *Malus florentina*

flore-pleno *FLOR-ee PLEE-no*
With double flowers, as in *Aquilegia vulgaris* var. *flore-pleno*

floribundus *flor-ih-BUN-dus*
floribunda, floribunium
floridus *flor-IH-dus*
florida, floridum
Very free-flowering, as in *Wisteria floribunda*

Pultenaea flexilis,
graceful bush pea

floridanus *flor-ih-DAH-nus*
floridana, floridanum
Connected with Florida, USA, as in *Illicium floridanum*

floriferus *flor-IH-fer-us*
florifera, floriferum
Very free-flowering, as in *Townsendia florifera*

flos *flos*
Used in combination to denote flower, as in *Lychnis flos-cuculi*
(cuckooflower)

fluitans *FLOO-ih-tanz*
Floating, as in *Glyceria fluitans*

fluminensis *floo-min-EN-sis*
fluminensis, fluminense
From Rio de Janeiro, Brazil, as in *Tradescantia fluminensis*

fluvialis *floo-vee-AHL-is*
fluvialis, fluviale
fluviatilis *floo-vee-uh-TIL-is*
fluviatilis, fluviatile
Growing in a river or running water, as in *Isotoma fluviatilis*

foeniculaceus *fen-ee-kul-ah-KEE-us*
foeniculacea, foeniculaceum
Like fennel (*Foeniculum*), as in *Argyranthemum foeniculaceum*

foetidus *FET-uh-dus*
foetida, foetidum
With a bad smell, as in *Vestia foetida*
foetidissimus *fet-uh-DISS-ih-mus*
foetidissima, foetidissimum
With a very bad smell, as in *Iris foetidissima*

foliaceus *foh-lee-uh-SEE-us*
foliacea, foliaceum
Like a leaf, as in *Aster foliaceus*

foliatus *fol-ee-AH-tus*
foliata, foliatum
With leaves, as in *Aletris foliata*

foliolotus *foh-lee-oh-LOH-tus*
foliolota, foliolotum
foliolosus *foh-lee-oh-LOH-sus*
foliolosa, foliolosum
With leaflets, as in *Thalictrum foliolosum*

FOENICULUM

Fennel is a good example of the confusion that can arise when a common name is loosely applied to two quite different plants from the same genus. *Foeniculum vulgare* var. *dulce* is the annual bulb-forming, low-growing plant often referred to as Florence fennel, the vegetable being something of a staple in Italian cooking. (*Vulgaris, vulgaris, vulgare* means common; *dulcis, dulcis, dulce* means sweet.) The tall and elegant perennial herb fennel is *F. vulgare* var. *sativum* and is grown for the culinary uses of its leaf and seed. Both green and bronze forms are available and, with its late-season umbelliferous flowers, it is such a beautiful plant that it appears as often in flower borders as in herb gardens. To avoid cross-pollination between the two varieties, ensure they are planted some distance apart. *F. vulgare* var. *sativum* is a prodigious self-seeder and the seedlings quickly develop a strong taproot, so should be weeded out as soon as they appear.

The Greek word for fennel is *marathon*. When the Persians were defeated at the Battle of Marathon in 490 BCE, a Greek runner was sent to run the 42 km (26 miles) to Athens to announce the news. The battle had been fought in a field of fennel (which type is unclear); thus *Foeniculum* and long-distance running have become forever linked. On a less heroic note, according to folklore, placing fennel seeds in keyholes is said to keep out ghosts.

Foeniculaceus (*foeniculacea foeniculaceum*) means resembling fennel. Several plants feature this term, including *Agastache foeniculum*, commonly known as anise hyssop, and

Both the leaves and seeds of *Foeniculum vulgare* have an aniseed taste. Technically the leaves are a herb and the seeds are a spice.

Lomatium foeniculaceum, the desert biscuitroot. However, the plant that counts fennel flower among its various common names, along with nutmeg flower and Roman coriander, is actually *Nigella hispanica*, which has no relation to the true fennels.

To swell and become succulent the bulbs need plenty of water.

If not given sufficient warmth *Foeniculum vulgare* var. *dulce*, Florence fennel, has a tendency to bolt into flower.

foliosus *foh-lee-OH-sus*
foliosa, foliosum
With many leaves; leafy, as in *Dactylorhiza foliosa*

follicularis *fol-lik-yoo-LAY-ris*
follicularis, folliculare
With follicles, as in *Cephalotus follicularis*

fontanus *FON-tah-nus*
fontana, fontanum
Growing in fast-running water, as in *Cerastium fontanum*

formosanus *for-MOH-sa-nus*
formosana, formosanum
Connected with Taiwan (formerly Formosa), as in *Pleione formosana*

formosus *for-MOH-sus*
formosa, formosum
Handsome; beautiful, as in *Pieris formosa*

forrestianus *for-rest-ee-AH-nus*
forrestiana, forrestianum
forrestii *for-rest-EE-eye*
Named after George Forrest (1873–1932), Scottish plant collector,
as in *Hypericum forrestii*

fortunei *for-TOO-nee-eye*
Named after Robert Fortune (1812–80), Scottish plant hunter and
horticulturist, as in *Trachycarpus fortunei*

foveolatus *foh-vee-oh-LAH-tus*
foveolata, foveolatum
With slight pitting, as in *Chionanthus foveolatus*

Pieris formosa

fragarioides *fray-gare-ee-OY-deez*
Resembling strawberry (*Fragaria*), as in *Waldsteinia fragarioides*

fragilis *FRAJ-ih-lis*
fragilis, fragile
Brittle; quick to wilt, as in *Salix fragilis*

fragrans *FRAY-granz*
Fragrant, as in *Osmanthus fragrans*

fragantissimus *fray-gran-TISS-ih-mus*
fragrantissima, fragrantissimum
Very fragrant, as in *Lonicera fragrantissima*

fraseri *FRAY-zer-ee*
Named after John Fraser (1750–1811), Scottish plant collector
and nurseryman, as in *Magnolia fraseri*

fraxineus *FRAK-si-nus*
fraxinea, fraxineum
Like ash (*Fraxinus*), as in *Blechnum fraxineum*

fraxinifolius *fraks-in-ee-FOH-lee-us*
fraxinifolia, fraxinifolium
With leaves like ash (*Fraxinus*), as in *Pterocarya fraxinifolia*

frigidus *FRIH-jih-dus*
frigida, frigidum
Growing in cold regions, as in *Artemisia frigida*

frondosus *frond-OH-sus*
frondosa, frondosum
Very leafy, as in *Primula frondosa*

frutescens *froo-TESS-enz*
fruticans *FROO-tih-kanz*
fruticosus *froo-tih-KOH-sus*
fruticosa, fruticosum
Shrubby; bushy, as in *Argyranthemum frutescens*

fruticola *froo-TIH-koh-luh*
Growing in bushy places, as in *Chirita fruticola*

fruticulosus *froo-tih-koh-LOH-sus*
fruticulosa, fruticulosum
Dwarf and shrubby, as in *Matthiola fruticulosa*

fucatus *few-KAH-tus*
fucata, fucatum
Painted; dyed, as in *Crocosmia fucata*

LATIN IN ACTION

Rubus fruticosus is a hardy spreading shrub and a firm favourite with soft-fruit growers (*fruticosus* meaning shrubby). For bumper crops of fruit, cut out the old fruited canes very soon after fruiting – this frees up space in which the new shoots can emerge and ripen before the winter months.

Rubus fruticosus,
blackberry

fuchsioides *few-shee-OY-deez*
Resembling *Fuchsia*, as in *Iochroma fuchsioides*

fugax *FOO-gaks*
Withering quickly; fleeting, as in *Urginea fugax*

fulgens *FUL-jenz*
fulgidus *FUL-jih-dus*
fulgida, fulgidum
Shining; glistening, as in *Rudbeckia fulgida*

fuliginosus *few-lih-gin-OH-sus*
fuliginosa, fuliginosum
A dirty brown or sooty colour, as in *Carex fuliginosa*

fulvescens *ful-VES-enz*
Becoming tawny in colour, as in *Masdevallia fulvescens*

fulvidus *FUL-vee-dus*
fulvida, fulvidum
Slightly tawny in colour, as in *Cortaderia fulvida*

fulvus *FUL-vus*
fulva, fulvum
Tawny orange in colour, as in *Hemerocallis fulva*

fumariifolius *foo-mar-ee-FOH-lee-us*
fumariifolia, fumariifolium
With leaves like fumitory (*Fumaria*), as in *Scabiosa fumariifolia*

funebris *fun-EE-bris*
funebris, funebre
Connected to graveyards, as in *Cupressus funebris*

fungosus *fun-GOH-sus*
fungosa, fungosum
Like fungus; spongy, as in *Borinda fungosa*

furcans *fur-kanz*
furcatus *fur-KA-tus*
furcata, furcatum
Forked, as in *Pandanus furcatus*

fuscatus *fus-KA-tus*
fuscata, fuscatum
Brownish in colour, as in *Sisyrinchium fuscatum*

fuscus *FUS-kus*
fusca, fuscum
A dusky or swarthy brown, as in *Nothofagus fusca*

futilis *FOO-tih-lis*
futilis, futile
Without use, as in *Salsola futilis*

G

gaditanus *gad-ee-TAH-nus*
gaditana, gaditanum
Connected with Cadiz, Spain, as in *Narcissus gaditanus*

galacifolius *guh-lay-sih-FOH-lee-us*
galacifolia, galacifolium
With leaves like wandplant (*Galax*), as in *Shortia galacifolia*

galanthus *guh-LAN-thus*
galantha, galanthum
With milky-white flowers, as in *Allium galanthum*

LATIN IN ACTION

Originally from Italian mountainous regions, the Adriatic bellflower is well suited to domestic cultivation in garden rockeries. Producing lovely star-shaped blue flowers in summer, it thrives in sun or light shade and a moist but well-drained soil, and creates a neat clump or mat reaching up to 10 cm (4 in).

Campanula garganica,
Adriatic bellflower

---

galeatus *ga-le-AH-tus*
galeata, galeatum
galericulatus *gal-er-ee-koo-LAH-tus*
galericulata, galericulatum
Shaped like a helmet, as in *Sparaxis galeata*

galegifolius *guh-lee-gih-FOH-lee-us*
galegifolia, galegifolium
With leaves like goat's rue (*Galega*), as in *Swainsona galegifolia*

gallicus *GAL-ih-kus*
gallica, gallicum
Connected with France, as in *Rosa gallica*

gangeticus *gan-GET-ih-kus*
gangetica, gangeticum
Of the Ganges regions of India and Bangladesh, as in *Asystasia gangetica*

garganicus *gar-GAN-ih-kus*
garganica, garganicum
Connected with Monte Gargano, Italy, as in *Campanula garganica*

gelidus *JEL-id-us*
gelida, gelidum
Connected with ice-cold regions, as in *Rhodiola gelida*

gemmatus *jem-AH-tus*
gemmata, gemmatum
Bejewelled, as in *Wikstroemia gemmata*

gemmiferus *jem-MIH-fer-us*
gemmifera, gemmiferum
With buds, as in *Primula gemmifera*

generalis *jen-er-RAH-lis*
generalis, generale
Normal, as in *Canna × generalis*

genevensis *gen-EE-ven-sis*
genevensis, genevense
From Geneva, Switzerland, as in *Ajuga genevensis*

geniculatus *gen-ik-yoo-LAH-tus*
geniculata, geniculatum
With a sharp bend like a knee, as in *Thalia geniculata*

genistifolius *jih-nis-tih-FOH-lee-us*
genistifolia, genistifolium
With leaves like broom (*Genista*), as in *Linaria genistifolia*

GERANIUM

The Latin genus name *Geranium* is so often used that it has more or less replaced the common name cranesbill; therefore you will often see the term geranium appearing in lower-case Roman type. The name derives from the Greek *geranos*, meaning crane, an allusion to the plant's elegant seed capsules, which look something like the long and pointed beak of a crane. These plants are so attractive, useful and easy to grow that it is not surprising that so many species have been introduced to gardens, all with tempting cultivars, making the gardener spoilt for choice when it comes to deciding which to grow.

Latin names come in very handy when faced with making a selection. For light shade, try *Geranium sylvaticum* (*sylvaticus, sylvatica, sylvaticum*, indicating it grows in woodlands), while *G. pratense* (*pratensis, pratensis, pratense*, meaning from the meadow) is the meadow cranesbill and has an informal, naturally relaxed habit with a greater spread than many other species. Another common name is herb robert. This is usually applied to *G. robertianum*, which is thought to have been named after a French abbot called Robert. Its other appellations include cuckoo's eye, death come quickly, red robin and stinking Robert.

The common term geranium is often casually but mistakenly interchanged with pelargonium. Although both belong to the family *Geraniaceae*, *Pelargonium* is a distinct genus from *Geranium*; it is not frost-hardy and is more suited to greenhouse cultivation or to planting out in

Pelargoniums make excellent ground cover, requiring little attention; they are happy in all but water-logged soils.

summer pots in cooler areas than its hardy cousin. It too derives its name from the Greek for a bird, this time a stork, *pelargos*, as the seed heads resemble the beak of that bird (its common name is storksbill). *Pelargonium peltatum* is often referred to as the ivy-leaved pelargonium as its leaves are shaped like a shield (*peltatus, peltata, peltatum*, relating to *pelta*, a shield), while *P. zonale* has distinctively marked leaves (*zonalis, zonalis, zonale*, meaning banded).

Pelargonium is not frost-hardy and is more suited to greenhouse cultivation.

LATIN IN ACTION

Bearded iris provide a startling range of colours in early summer. Plant in a well-drained site with the rhizomes sitting just under the surface of the soil and the roots spread out. If congested, the clumps should be divided and replanted after flowering. This can be done every few years to ensure the plants continue to bloom well. The flowers have soft hairs on their lower petals, known as the beard.

Iris germanica, bearded iris

geoides *jee-OY-deez*
Like avens (*Geum*), as in *Waldsteinia geoides*

geometrizans *jee-oh-MET-rih-zanz*
With markings in a formal pattern, as in *Myrtillocactus geometrizans*

georgianus *jorj-ee-AH-nus*
georgiana, georgianum
Connected with Georgia, USA, as in *Quercus georgiana*

georgicus *JORJ-ih-kus*
georgica, georgicum
Connected with Georgia (Eurasia), as in *Pulsatilla georgica*

geranioides *jer-an-ee-OY-deez*
Resembling *Geranium*, as in *Saxifraga geranioides*

germanicus *jer-MAN-ih-kus*
germanica, germanicum
Connected with Germany, as in *Iris germanica*

gibberosus *gib-er-OH-sus*
gibberosa, gibberosum
With a hump on one side, as in *Scaphosepalum gibberosum*

gibbiflorus *gib-bih-FLOR-us*
gibbiflora, gibbiflorum
With flowers that have a hump on one side, as in *Echeveria gibbiflora*

gibbosus *gib-OH-sus*
gibbosa, gibbosum
gibbus *gib-us*
gibba, gibbum
With a swelling on one side, as in *Fritillaria gibbosa*

gibraltaricus *jib-ral-TAH-rih-kus*
gibraltarica, gibraltaricum
Connected with Gibraltar, Europe, as in *Iberis gibraltarica*

giganteus *jy-GAN-tee-us*
gigantea, giganteum
Unusually tall or large, as in *Stipa gigantea*

giganthus *jy-GAN-thus*
gigantha, giganthum
With large flowers, as in *Hemsleya gigantha*

gilvus *GIL-vus*
gilva, gilvum
Dull yellow, as in *Echeveria × gilva*

glabellus *gla-BELL-us*
glabella, glabellum
Smooth, as in *Epilobium glabellum*

glaber *glay-ber*
glabra, glabrum
Smooth, hairless, as in *Bougainvillea glabra*

glabratus *GLAB-rah-tus*
glabrata, glabratum
glabrescens *gla-BRES-senz*
glabriusculus *gla-bree-US-kyoo-lus*
glabriuscula, glabriusculum
Rather hairless, as in *Corylopsis glabrescens*

glacialis *glass-ee-AH-lis*
glacialis, glaciale
Connected with ice-cold, glacial regions, as in
Dianthus glacialis

gladiatus *glad-ee-AH-tus*
gladiata, gladiatum
Like a sword, as in *Coreopsis gladiata*

glanduliferus *glan-doo-LIH-fer-us*
glandulifera, glanduliferum
With glands, as in *Impatiens glandulifera*

glanduliflorus *gland-yoo-LIH-flor-us*
glanduliflora, glanduliflorum
With glandular flowers, as in *Stapelia glanduliflora*

glandulosus *glan-doo-LOH-sus*
glandulosa, glandulosum
Glandular, as in *Erodium glandulosum*

glaucescens *glaw-KES-enz*
With a bloom; blue-green in colour, as in *Ferocactus glaucescens*

glaucifolius *glau-see-FOH-lee-us*
glaucifolia, glaucifolium
With grey-green leaves; with leaves with a bloom, as in
Diospyros glaucifolia

glaucophyllus *glaw-koh-FIL-us*
glaucophylla, glaucophyllum
With grey-green leaves, or with a bloom, as in
Rhododendron glaucophyllum

glaucus *GLAW-kus*
glauca, glaucum
With a bloom on the leaves, as in *Festuca glauca*

Artemisia glacialis,
Alpine mugwort

globiferus *glo-BIH-fer-us*
globifera, globiferum
With spherical clusters of small globes, as in *Pilularia globifera*

globosus *glo-BOH-sus*
globosa, globosum
Round, as in *Buddleja globosa*

globuliferus *glob-yoo-LIH-fer-us*
globulifera, globuliferum
With small spherical clusters, as in *Saxifraga globulifera*

globularis *glob-YOO-lah-ris*
globularis, globulare
Relating to a small sphere, as in *Carex globularis*

globuligemma *glob-yoo-lih-JEM-uh*
With round buds, as in *Aloe globuligemma*

globulosus *glob-yoo-LOH-sus*
globulosa, globulosum
Small and spherical, as in *Hoya globulosa*

JOHN BARTRAM
(1699–1777)

WILLIAM BARTRAM
(1739–1823)

From the mid-1730s onwards, the Linnaean approach of careful and systematic plant collection, identification and recording began to be practised by North American botanists and naturalists. The remarkable self-taught botanist John Bartram was at the forefront of the promotion of this rigorous system. A modest farmer and landowner with little formal education, the garden he created at his home Kingsessing, near Philadelphia, is now considered to be the country's first botanical garden. The Bartrams were Quakers and the Society of Friends in the Philadelphia area was very active in promoting the study of plants and horticulture during this period.

Bartram's influence spread abroad with surprising rapidity and on an unprecedented scale; it was his supply of native seed to the English wool merchant Peter Collinson that resulted in around 200 new trees, shrubs and plants of American origin being introduced into Britain. Throughout his career he sent numerous specimens and seeds to noted botanists in Europe, including to London's Chelsea Physic Garden and Kew Gardens. From 1735 onwards, he travelled extensively in the southeastern states, collecting and recording what he found with accuracy and clarity. Bartram was introduced to Pehr Kalm, a Finnish former student of Carl Linnaeus (see p. 132), and a dialogue of scientific discoveries was quickly established between Bartram and Linnaeus. Such was the fame of his unique collection of wild plants that the English king, George III, appointed him to the post of King's Botanist in April 1765 at an annual fee of £50; by the end of the decade, Bartram had been elected to the Royal Academy of Science in Stockholm. Among his many discoveries are the cucumber tree, *Magnolia acuminata*, the tupelo, *Nyssa sylvatica*, and the Venus fly trap, *Dionaea muscipula*.

John Bartram engrossed in botanising.

Note the flowers adorning the attire of William Bartram in this portrait by Charles Willson Peale.

As if one important botanist in a family were not enough, John Bartram's fifth child, William, is still revered today as an important naturalist, explorer and man of letters. As a young man he accompanied his father on many expeditions, including the 1765–6 exploration of 400 miles of the St. John's River. After a thwarted attempt to become a plantation owner growing indigo in Florida, he returned to Philadelphia, then set out on what was to become a four-year-long solitary journey through the southern colonies. The resulting volume, *Travels Through North and South Carolina, Georgia, East and West Florida, the Cherokee Country etc.*, published in 1791, was a clear and sensitive account of the geography and natural history of these regions, and offered new insights into the lives and customs of the indigenous people who lived there. He was also a great authority on the native birds of the United States, and a genus of birds, *Bartramia*, is named after him. Due to ill health he was unable to accept the post of Professor of Botany at the University of Pennsylvania. Likewise, his poor eyesight prevented him from accepting an offer to join Lewis and Clark's expedition (see p. 54) as Official Naturalist.

The highly ornamental Franklin tree, *Franklinia alatamaha*, was discovered by John Bartram on his 1765 trip, which took him down the Cape Fear and Altamaha Rivers in Georgia. He named it in honour of his friend, Benjamin Franklin.

William Bartram's empathetic viewpoint was particularly appreciated in Europe, where he was widely read. In the 19th century, he was to influence the English Romantic poets William Wordsworth and Samuel Taylor Coleridge, while back in the United States, Ralph Waldo Emerson and Henry David Thoreau were among his admirers. As well as writing beautiful poetic prose, William Bartram was a talented artist and he recorded with great accuracy and fluency the flora and fauna encountered on his travels. Many of his surviving paintings and drawings are housed in London's Natural History Museum. His evident sensitivity to the natural world chimes with modern environmental concerns and his works are still enjoyed today.

William and his brother John Bartram Jr. continued to develop the garden their father had established and to maintain the thriving family business that so successfully exported North American plants and seeds to many countries around the world. Plants named in the family's honour include *Amelanchier bartramiana* and *Commersonia bartramia*. There is also a genus of mosses named *Bartramia* (from the family *Bartramiaceae*). Ten years after John Bartram's death, George Washington visited and admired his impressive botanical garden at Kingsessing, and it has continued to welcome visitors ever since.

'THE GREATEST BOTANIST IN THE WORLD.'
Carl Linnaeus (1707–78), speaking about John Bartram

glomeratus *glom-er-AH-tus*
glomerata, glomeratum
With clusters of rounded heads, as in *Campanula glomerata*

gloriosus *glo-ree-OH-sus*
gloriosa, gloriosum
Superb; glorious, as in *Yucca gloriosa*

gloxinioides *gloks-in-ee-OY-deez*
Resembling *Gloxinia*, as in *Penstemon gloxinioides*

glumaceus *gloo-MA-see-us*
glumacea, glumaceum
With glumes (the bracts that enclose the flowers of grasses),
as in *Dendrochilum glumaceum*

glutinosus *gloo-tin-OH-sus*
glutinosa, glutinosum
Sticky; glutinous, as in *Eucryphia glutinosa*

glycinoides *gly-sin-OY-deez*
Like soybean (*Glycine*), as in *Clematis glycinoides*

gnaphaloides *naf-fal-OY-deez*
Like cudweed (*Gnaphalium*), as in *Senecio gnaphaloides*

gongylodes *GON-jih-loh-deez*
Swollen; roundish, as in *Cissus gongylodes*

goniocalyx *gon-ee-oh-KAL-iks*
Calyx with corners or angles, as in *Eucalyptus goniocalyx*

gossypinus *goss-ee-PEE-nus*
gossypina, gossypinum
Like cotton (*Gossypium*), as in *Strobilanthes gossypina*

gracilentus *grass-il-EN-tus*
gracilenta, gracilentum
Graceful; slender, as in *Rhododendron gracilentum*

graciliflorus *grass-il-ih-FLOR-us*
graciliflora, graciliflorum
With slender or graceful flowers, as in *Pseuderanthemum
graciliflorum*

gracilipes *gra-SIL-i-peez*
With a slender stalk, as in *Mahonia gracilipes*

gracilis *GRASS-il-is*
gracilis, gracile
Graceful; slender, as in *Geranium gracile*

Lathyrus grandiflorus,
perennial sweet pea

graecus *GRAY-kus*
graeca, graecum
Greek, as in *Fritillaria graeca*

gramineus *gram-IN-ee-us*
graminea, gramineum
Like grass, as in *Iris graminea*

graminifolius *gram-in-ee-FOH-lee-us*
graminifolia, graminifolium
With grass-like leaves, as in *Stylidium graminifolium*

granadensis *gran-uh-DEN-sis*
granadensis, granadense
From Granada, Spain or Colombia, South America,
as in *Drimys granadensis*

grandis *gran-DIS*
grandis, grande
Big; showy, as in *Licuala grandis*

grandiceps *GRAN-dee-keps*
With large head, as in *Leucogenes grandiceps*

grandicuspis *gran-dih-KUS-pis*
grandicuspis, grandicuspe
With big points, as in *Sansevieria grandicuspis*

grandidentatus *gran-dee-den-TAH-tus*
grandidentata, grandidentatum
With big teeth, as in *Thalictrum grandidentatum*

grandiflorus *gran-dih-FLOR-us*
grandiflora, grandiflorum
With large flowers, as in *Platycodon grandiflorus*

grandifolius *gran-dih-FOH-lee-us*
grandifolia, grandifolium.
With large leaves, as in *Haemanthus grandifolius*

graniticus *gran-NY-tih-kus*
granitica, graniticum
Growing on granite and rocks, as in *Dianthus graniticus*

granulatus *gran-yoo-LAH-tus*
granulata, granulatum
Bearing grain-like structures, as in *Saxifraga granulata*

granulosus *gran-yool-OH-sus*
granulosa, granulosum
Made of small grains, as in *Centropogon granulosus*

gratianopolitanus *grat-ee-an-oh-pol-it-AH-nus*
gratianopolitana, gratianopolitanum
Connected with Grenoble, France, as in *Dianthus gratianopolitanus*

gratissimus *gra-TIS-ih-mus*
gratissima, gratissimum
Very pleasing, as in *Luculia gratissima*

gratus *GRAH-tus*
grata, gratum
Giving pleasure, as in *Conophytum gratum*

graveolens *grav-ee-OH-lenz*
With a heavy scent, as in *Ruta graveolens*

griseus *GREE-see-us*
grisea, griseum
Grey, as in *Acer griseum*

grosseserratus *grose ser-AH-tus*
grosseserrata, grosseserratum
With large saw teeth, as in *Clematis occidentalis* subsp. *grosseserrata*

grossus *GROSS-us*
grossa, grossum
Very large, as in *Betula grossa*

guianensis *gee-uh-NEN-sis*
guianensis, guianense
From Guiana, South America, as in *Couroupita guianensis*

guineensis *gin-ee-EN-sis*
guineensis, guineense
From the Guinea coast, West Africa, as in *Elaeis guineensis*

gummifer *GUM-mif-er*
gummifera, gummiferum
Producing gum, as in *Seseli gummiferum*

gummosus *gum-MOH-sus*
gummosa, gummosum
Gummy, as in *Ferula gummosa*

guttatus *goo-TAH-tus*
guttata, guttatum
With spots, as in *Mimulus guttatus*

gymnocarpus *jim-noh-KAR-pus*
gymnocarpa, gymnocarpum
With naked, uncovered fruit, as in *Rosa gymnocarpa*

H

haastii *HAAS-tee-eye*
Named after Sir Julius von Haast (1824–87), German explorer
and geologist, as in *Olearia × haastii*

hadriaticus *had-ree-AT-ih-kus*
hadriatica, hadriaticum
Connected with the shores of the Adriatic Sea, Europe, as in
Crocus hadriaticus

haemanthus *hem-AN-thus*
haemantha, haemanthum
With blood-red flowers, as in *Alstroemeria haemantha*

haematocalyx *hem-at-oh-KAL-icks*
With a blood-red calyx, as in *Dianthus haematocalyx*

haematochilus *hem-mat-oh-KY-lus*
haematochila, haematochilum
With a blood-red lip, as in *Oncidium haematochilum*

Glechoma hederacea,
ground ivy

haematodes *hem-uh-TOH-deez*
Blood-red, as in *Rhododendron haematodes*

hakeoides *hak-ee-OY-deez*
Resembles *Hakea*, as in *Berberis hakeoides*

halophilus *hal-oh-FIL-ee-us*
halophila, halophilum
Salt-loving, as in *Iris spuria* subsp. *halophila*

hamatus *ham-AH-tus*
hamata, hamatum
hamosus *ham-UH-sus*
hamosa, hamosum
Hooked, as in *Euphorbia hamata*

harpophyllus *harp-oh-FIL-us*
harpophylla, harpophyllum
With leaves shaped like sickles, as in *Laelia harpophylla*

hastatus *hass-TAH-tus*
hastata, hastatum
Shaped like a spear, as in *Verbena hastata*

hastilabius *hass-tih-LAH-bee-us*
hastilabia, hastilabium
With a spear-shaped lip, as in *Oncidium hastilabium*

hastulatus *hass-TOO-lat-tus*
hastulata, hastulatum
Shaped somewhat like a spear, as in *Acacia hastulata*

hebecarpus *hee-be-KAR-pus*
hebecarpa, hebecarpum
With down-covered fruit, as in *Senna hebecarpa*

hebephyllus *hee-bee-FIL-us*
hebephylla, hebephyllum
With down-covered leaves, as in *Cotoneaster hebephyllus*

hederaceus *hed-er-AYE-see-us*
hederacea, hederaceum
Like ivy (*Hedera*), as in *Glechoma hederacea*

hederifolius *hed-er-ih-FOH-lee-us*
hederifolia, hederifolium
With leaves like ivy (*Hedera*), as in *Veronica hederifolia*

helianthoides *hel-ih-anth-OH-deez*
Resembling sunflower (*Helianthus*), as in *Heliopsis helianthoides*

HELIANTHUS

With its big and bold flowers, it is perhaps not too surprising that the joyous sunflower has an enduring place in the customs and mythologies of many cultures. Peruvian and Inca civilisations worshipped the sunflower as a symbol of the sun, and Native Americans placed sunflower seeds on the graves of their dead. For the Chinese, the flower is a symbol of longevity, while European folklore advises picking a sunflower at sunset while making a wish then, with the rising of the sun the next day, the wish will be granted.

As so often with plant names, its Latin nomenclature has Greek roots. *Helianthus* comes from the Greek for sun, *helios*, and *anthos*, meaning flower. Belonging to the *Asteraceae* family, the floriferous *Helianthus annuus* should not be confused with *H. tuberosus*, which is the vegetable Jerusalem artichoke (interestingly, another common name of this plant is 'sunchoke'). Species names include *H. decapetalus*, commonly referred to as forest sunflower or thinleaf sunflower (*decapetalus, decapetala, decapetalum*, with ten petals). As its subspecies name suggests, *H. debilis* subsp. *cucumerifolius* has leaves resembling the cucumber plant (*debilis, debilis, debile*, means weak and frail).

A further linguistic association with the sun is 'rays', the correct botanical term used for the petal-like parts of daisy flowers. Other plant genera with *helio* in their name include *Heliopsis* and *Heliotropium*. The term heliotropic describes the habit of some plants whereby their stalks and flowers follow the movement of the sun across the sky during the course of a day (*trope* is Greek for turning).

Along with the ornamental qualities of their striking flowers, sunflowers are also grown as a crop as they produce excellent oil and their seeds are edible and highly nutritious. This group of annuals and perennials is easy to grow from seed; the main cultivation challenge is giving them sufficient water and a strong enough stake for support. Their tendency to grow and grow is legendary, with heights of 8 m (26 ft) or more being recorded.

Children love seeing the small seeds of the sunflower turn into a towering bloom much taller than themselves.

Helianthus annuus,
sunflower

helix *HEE-licks*

Spiral-shaped; applied to twining plants, as in *Hedera helix*

hellenicus *hel-LEN-ih-kus*

hellenica, hellenicum

Connected with Greece, as in *Linaria hellenica*

helodes *hel-OH-deez*

From boggy land, as in *Drosera helodes*

LATIN IN ACTION

English or common ivy is the least demanding of plants. Evergreen, with a twining and clinging habit (*helix* means spiral-shaped), it will grow in sun or shade and is fully hardy. In winter its berries provide an important food source for birds, while in late summer and autumn bees feast on its small white flowers.

Hedera helix,
English ivy

---

helveticus *hel-VET-ih-kus*

helvetica, helveticum

Connected with Switzerland, as in *Erysimum helveticum*

helvolus *HEL-vol-us*

helvola, helvolum

Reddish yellow, as in *Vanda helvola*

hemisphaericus *hem-is-FEER-ih-kus*

hemisphaerica, hemisphaericum

In the shape of half a sphere, as in *Quercus hemisphaerica*

henryi *HEN-ree-eye*

Named after Augustine Henry (1857–1930), Irish plant collector, as in *Lilium henryi*

hepaticifolius *hep-at-ih-sih-FOH-lee-us*

hepaticifolia, hepaticifolium

With leaves like liverwort (*Hepatica*), as in *Cymbalaria hepaticifolia*

hepaticus *hep-AT-ih-kus*

hepatica, hepaticum

Dull brown; the colour of liver, as in *Anemone hepatica*

hepta-

Used in compound words to denote seven

heptaphyllus *hep-tah-FIL-us*

heptaphylla, heptaphyllum

With seven leaves, as in *Parthenocissus heptaphylla*

heracleifolius *hair-uh-klee-ih-FOH-lee-us*

heracleifolia, heracleifolium

With leaves like hogweed (*Heracleum*), as in *Begonia heracleifolia*

herbaceus *her-buh-KEE-us*

herbacea, herbaceum

Herbaceous, i.e. not woody, as in *Salix herbacea*

heter-, hetero-

Used in compound words to denote various or diverse

heteracanthus *het-er-a-KAN-thus*

heteracantha, heteracanthum

With various or diverse spines, as in *Agave heteracantha*

heteranthus *het-er-AN-thus*

heterantha, heteranthum

With various or diverse flowers, as in *Indigofera heterantha*

heterocarpus *het-er-oh-KAR-pus*
heterocarpa, heterocarpum
With various or diverse fruit, as in *Ceratocapnos heterocarpa*

heterodoxus *het-er-oh-DOKS-us*
heterodoxa, heterodoxum
Differing from the type of the genus, as in *Heliamphora heterodoxa*

heteropetalus *het-er-oh-PET-uh-lus*
heteropetala, heteropetalum
With various or diverse petals, as in *Erepsia heteropetala*

heterophyllus *het-er-oh-FIL-us*
heterophylla, heterophyllum
With various or diverse leaves, as in *Osmanthus heterophyllus*

heteropodus *het-er-oh-PO-dus*
heteropoda, heteropodum
With various or diverse stalks, as in *Berberis heteropoda*

hexa-
Used in compound words to denote six

hexagonopterus *heks-uh-gon-OP-ter-us*
hexagonoptera, hexagonopterum
With six-angled wings, as in *Phegopteris hexagonoptera*

hexagonus *hek-sa-GON-us*
hexagona, hexagonum
With six angles, as in *Cereus hexagonus*

hexandrus *heks-AN-drus*
hexandra, hexandrum
With six stamens, as in *Sinopodophyllum hexandrum*

hexapetalus *heks-uh-PET-uh-lus*
hexapetala, hexapetalum
With six petals, as in *Ludwigia grandiflora*
subsp. *hexapetala*

hexaphyllus *heks-uh-FIL-us*
hexaphylla, hexaphyllum
With six leaves or leaflets, as in *Stauntonia hexaphylla*

hians *HY-anz*
Gaping, as in *Aeschynanthus hians*

hibernicus *hy-BER-nih-kus*
hibernica, hibernicum
Connected with Ireland, as in *Hedera hibernica*

hiemalis *hy-EH-mah-lis*
hiemalis, hiemale
Of the winter; winter-flowering, as in *Leucojum hiemale*

hierochunticus *hi-er-oh-CHUN-tih-kus*
hierochuntica, hierochunticum
Connected with Jericho, as in *Anastatica hierochuntica*

himalaicus *him-al-LAY-ih-kus*
himalaica, himalaicum
Connected with the Himalaya, as in *Stachyurus himalaicus*

himalayensis *him-uh-lay-EN-is*
himalayensis, himalayense
From the Himalaya, as in *Geranium himalayense*

hircinus *her-SEE-nus*
hircina, hircinum
Goat-like, or with a goat-like odour, as in *Hypericum hircinum*

Penstemon heterophyllus,
foothill penstemon

hirsutus *her-SOO-tus*
hirsuta, hirsutum
Hairy, as in *Lotus hirsutus*

hirsutissimus *her-soot-TEE-sih-mus*
hirsutissima, hirsutissimum
Very hairy, as in *Clematis hirsutissima*

hirsutulus *her-SOOT-oo-lus*
hirsutula, hirsutulum
Somewhat hairy, as in *Viola hirsutula*

hirtellus *her-TELL-us*
hirtella, hirtellum
Rather hairy, as in *Plectranthus hirtellus*

Narcissus hispanicus,
Spanish daffodil

hirtiflorus *her-tih-FLOR-us*
hirtiflora, hirtiflorum
With hairy flowers, as in *Passiflora hirtiflora*

hirtipes *her-TYE-pees*
With hairy stems, as in *Viola hirtipes*

hirtus *HER-tus*
hirta, hirtum
Hairy, as in *Columnea hirta*

hispanicus *his-PAN-ih-kus*
hispanica, hispanicum
Connected with Spain, as in *Narcissus hispanicus*

hispidus *HISS-pih-dus*
hispida, hispidum
With bristles, as in *Leontodon hispidus*

hollandicus *hol-LAN-dih-kus*
hollandica, hollandicum
Connected with Holland, as in *Allium hollandicum*

holo-
Used in compound words to denote completely

holocarpus *ho-loh-KAR-pus*
holocarpa, holocarpum
With complete fruit, as in *Staphylea holocarpa*

holochrysus *ho-loh-KRIS-us*
holochrysa, holochrysum
Completely golden, as in *Aeonium holochrysum*

holosericeus *ho-loh-ser-ee-KEE-us*
holosericea, holosericeum
With silky hairs all over, as in *Convolvulus holosericeus*

horizontalis *hor-ih-ZON-tah-lis*
horizontalis, horizontale
Close to the ground; horizontal, as in *Cotoneaster horizontalis*

horridus *HOR-id-us*
horrida, horridum
With many prickles, as in *Euphorbia horrida*

hortensis *hor-TEN-sis*
hortensis, hortense
hortorum *hort-OR-rum*
hortulanus *hor-tew-LAH-nus*
hortulana, hortulanum
Relating to gardens, as in *Lysichiton × hortensis*

As its name indicates, *Eranthis hyemalis* flowers in winter, and few sights are so welcome as a yellow and white carpet of winter aconites and early snowdrops. They do best in full sun or dappled shade and should be planted en masse for greatest impact.

Eranthis hyemalis,
winter aconite

hugonis *hew-GO-nis*
Named after Father Hugh Scallon, missionary in China in the late 19th and early 20th centuries, as in *Rosa hugonis*

humifusus *hew-mih-FEW-sus*
humifusa, humifusum
Of a sprawling habit, as in *Opuntia humifusa*

humilis *HEW-mil-is*
humilis, humile
Low-growing, dwarfish, as in *Chamaerops humilis*

hungaricus *hun-GAR-ih-kus*
hungarica, hungaricum
Connected with Hungary, as in *Colchicum hungaricum*

hunnewellianus *hun-ee-we-el-AH-nus*
hunnewelliana, hunnewellianum
Named after the Hunnewell family of the Hunnewell Arboretum, Wellesley, USA, as in *Rhododendron hunnewellianum*

hupehensis *hew-pay-EN-sis*
hupehensis, hupehense
From Hupeh (Hubei), China, as in *Sorbus hupehensis*

hyacinthinus *hy-uh-sin-THEE-nus*
hyacinthina, hyacinthinum
hyacinthus *hy-uh-SIN-thus*
hyacintha, hyacinthum
Dark purple-blue, or like a hyacinth, as in *Triteleia hyacinthina*

hyalinus *hy-yuh-LEE-nus*
hyalina, hyalinum
Transparent; almost transparent, as in *Allium hyalinum*

hybridus *hy-BRID-us*
hybrida, hybridum
Mixed; hybrid, as in *Helleborus × hybridus*

hydrangeoides *hy-drain-jee-OY-deez*
Resembling *Hydrangea*, as in *Schizophragma hydrangeoides*

hylaeus *hy-la-ee-us*
hylaea, hylaeum
From the woods, as in *Rhododendron hylaeum*

hyemalis *hy-EH-mah-lis*
hyemalis, hyemale
Relating to winter; winter-flowering, as in *Eranthis hyemalis*

hymen-
Used in compound words to denote membranous

hymenanthus *hy-men-AN-thus*
hymenantha, hymenanthum
With flowers with a membrane, as in *Trichopilia hymenantha*

hymenorrhizus *hy-men-oh-RY-zus*
hymenorrhiza, hymenorrhizum
With membranous roots, as in *Allium hymenorrhizum*

hymenosepalus *hy-men-no-SEP-uh-lus*
hymenosepala, hymenosepalum
With membranous sepals, as in *Rumex hymenosepalus*

hyperboreus *hy-puh-BOR-ee-us*
hyperborea, hyperboreum
Connected with the far north, as in *Sparganium hyperboreum*

hypericifolius *hy-PER-ee-see-FOH-lee-us*
hypericifolia, hypericifolium
With leaves like St. John's wort (*Hypericum*), as in
Melalaeuca hypericifolia

hypericoides *hy-per-ih-KOY-deez*
Resembling *Hypericum*, as in *Ascyrum hypericoides*

hypnoides *hip-NO-deez*
Resembling moss, as in *Saxifraga hypnoides*

Hypericum perforatum,
perforate St John's wort

hypo-
Used in compound words to denote under

hypochondriacus *hy-po-kon-dree-AH-kus*
hypochondriaca, hypochondriacum
With a melancholy appearance; with dull-coloured flowers, as in
Amaranthus hypochondriacus

hypogaeus *hy-poh-JEE-us*
hypogaea, hypogaeum
Underground; developing in the earth, as in *Copiapoa hypogaea*

hypoglaucus *hy-poh-GLAW-kus*
hypoglauca, hypoglaucum
Glaucous underneath, as in *Cissus hypoglauca*

hypoglottis *hh-poh-GLOT-tis*
Underside of the tongue, from the shape of the pods, as in
Astragalus hypoglottis

hypoleucus *hy-poh-LOO-kus*
hypoleuca, hypoleucum
White underneath, as in *Centaurea hypoleuca*

hypophyllus *hy-poh-FIL-us*
hypophylla, hypophyllum
Underneath the leaf, as in *Ruscus hypophyllum*

hypopitys *hi-po-PY-tees*
Growing under pines, as in *Monotropa hypopitys*

hyrcanus *hyr-KAH-nus*
hyrcana, hyrcanum
Connected with the region of the Caspian Sea (Hyrcania in
antiquity), as in *Hedysarum hyrcanum*

hyssopifolius *hiss-sop-ih-FOH-lee-us*
hyssopifolia, hyssopifolium
With leaves like hyssop (*Hyssopus*), as in *Cuphea hyssopifolia*

hystrix *HIS-triks*
Bristly; like a porcupine, as in *Colletia hystrix*

I

ibericus *eye-BEER-ih-kus*
iberica, ibericum
Connected with Iberia (Spain and Portugal), as in
Geranium ibericum

iberidifolius *eye-beer-id-ih-FOH-lee-us*
iberidifolia, iberidifolium
With leaves that resemble *Iberis*, as in *Brachyscome iberidifolia*

icos-
Used in compound words to denote twenty

icosandrus *eye-koh-SAN-drus*
icosandra, icosandrum
With twenty stamens, as in *Phytolacca icosandra*

idaeus *eye-DAY-ee-us*
idaea, idaeum
Connected with Mount Ida, Crete, as in *Rubus idaeus*

ignescens *ig-NES-enz*
igneus *ig-NE-us*
ignea, igneum
Fiery red, as in *Cuphea ignea*

ikariae *eye-KAY-ree-ay*
Of the island of Ikaria, in the Aegean Sea, as in *Galanthus ikariae*

ilicifolius *il-liss-ee-FOH-lee-us*
ilicifolia, ilicifolium
With leaves like holly (*Ilex*), as in *Itea ilicifolia*

illecebrosus *il-lee-see-BROH-sus*
illecebrosa, illecebrosum
Enticing; charming, as in *Tigridia illecebrosa*

illinitus *il-lin-EYE-tus*
illinita, illinitum
Smeared; smirched, as in *Escallonia illinita*

illinoinensis *il-ih-no-in-EN-sis*
illinoinensis, illinoinense
From Illinois, USA, as in *Carya illinoinensis*

illustris *il-LUS-tris*
illustris, illustre
Brilliant; lustrous, as in *Amsonia illustris*

Crocus imperati

illyricus *il-LEER-ih-kus*
illyrica, illyricum
Connected with Illyria, the name for an area of the western Balkan
Peninsula in antiquity, as in *Gladiolus illyricus*

ilvensis *il-VEN-sis*
ilvensis, ilvense
Of Elba, Italy, or the River Elbe, as in *Woodsia ilvensis*

imberbis *IM-ber-bis*
imberbis, imberbe
Without spines or beard, as in *Rhododendron imberbe*

imbricans *IM-brih-KANS*
imbricatus *IM-brih-KA-tus*
imbricata, imbricatum
With elements that overlap in a regular pattern, as in
Gladiolus imbricatus

immaculatus *im-mak-yoo-LAH-tus*
immaculata, immaculatum
Spotless, as in *Aloe immaculata*

immersus *im-MER-sus*
immersa, immersum
Growing under water, as in *Pleurothallis immersa*

imperati *im-per-AH-tee*
Named after Ferrante Imperato (1550–1625), an apothecary from
Naples, Italy, as in *Crocus imperati*

DAVID DOUGLAS

(1799–1834)

David Douglas is the most famous botanical explorer of North America. His influence is important as it was his introduction of particular trees and plants that shaped the appearance of so many of the great gardens of North America and Europe. Born in Scone, Scotland, he was apprenticed from the age of 11 to the head gardener at Scone Palace, the Earl of Mansfield's estate. He later worked in several other Scottish gardens and thus gained access to some fine botanical libraries, in which he assiduously pursued his study of botany. It was while he was at the Glasgow Botanical Gardens that the then Professor of Botany, William Hooker (see p. 182), taught the young Douglas to collect, identify and press botanical specimens.

Due to the fulsome recommendations of Hooker, the Horticultural Society of London (later the Royal Horticultural Society) decided to send Douglas on a plant-collecting expedition to the northeast coast of the United States in 1823. During this trip, he visited New York, Philadelphia, Lake Eire, Buffalo and Niagara Falls. He also visited the botanical garden of the late John Bartram (see p. 98). Douglas returned to England with numerous ornamental plants and trees, including several new fruiting cultivars. So successful was his first expedition considered that he was soon sent across the Atlantic again, this time to the Pacific Northwest. The journey was primarily sponsored by the Horticultural Society of London, as well as the Hudson's Bay Company and he collected plants and other natural history objects of interest along the Columbia River. During his three-year stay, he encountered the massive *Pseudotsuga menziesii*, the tree that now bears the common name Douglas fir. Of this he wrote that it 'exceeds all trees in magnitude. I measured one lying on the shores of the river 39 feet in circumference and 159 feet long; the top was wanting ... so I judge that it would be in all about 190 feet high.'

On this trip he also found the giant tree *Pinus lambertiana*. Due to the difficulties presented by its extreme height, he struck on an innovative method of harvesting the cones – he fired his gun at the upper branches and down the cones fell! Unfortunately, the noise alerted some rather aggressive neighbouring Native Americans and he was forced to make a quick getaway. Douglas sent back to England more than 500 specimens of plants, along with seeds and bird and animal skins. During his third expedition to the United States, he travelled to California and claimed to have identified 20 genera and 360 species hitherto unknown to science, among them the noble fir, *Abies procera*.

During his career, Douglas collected literally hundreds of trees, shrubs, ornamental plants, herbs

Despite little formal education, Douglas gained a high level of botanical knowledge and his legacy is still evident in our public parks and private gardens.

'ENDOWED WITH AN ACUTE AND VIGOROUS MIND
WHICH HE IMPROVED BY DILIGENT STUDY.'

Memorial inscription, Old Parish Church, Scone, Scotland

Pseudotsuga menziesii,
Douglas fir

The eponymous Douglas fir, *Pseudotsuga menziesii*,
was first introduced into cultivation by David Douglas,
although it had been identified by Lewis and Clark on
their earlier expedition.

Crataegus douglasii,
Douglas' thornapple

Douglas collected seeds of this bushy shrub. It is
also known as black haw and Douglas' hawthorn.

and mosses. His North American collection,
including that of northern Canada by Sir John
Richardson and Thomas Drummond, was catalogued
in Hooker's *Flora boreali-americana*, published in
two volumes between 1829 and 1840. *Crataegus
douglasii, Juniperus horizontalis* 'Douglasii' and
Quercus douglasii are among the plants named in his
honour. His journeys were often hazardous and rarely
without incident; dramas included bolting horses,
lost collections, upturned canoes, stolen belongings

and storm-tossed vessels. However, it was his 1834
trip to the Sandwich Islands (the Hawaiian Islands)
that was to bring the greatest disaster of all. Although
he was only 35, Douglas's eyesight was failing, and
this may have contributed to his accidental fall into a
pit trap. Such cavernous holes were dug to trap large
animals and it is thought that the one Douglas fell
into may have already contained a bullock. His body
was later found by missionaries, horribly gored and
with his faithful dog still waiting by the pit's edge.

imperialis *im-peer-ee-AH-lis*
imperialis, imperiale
Very fine; showy, as in *Fritillaria imperialis*

implexus *im-PLECK-sus*
implexa, implexum
Tangled, as in *Kleinia implexa*

impressus *im-PRESS-us*
impressa, impressum
With impressed or sunken surfaces, as in *Ceanothus impressus*

inaequalis *in-ee-KWA-lis*
inaequalis, inaequale
Unequal, as in *Geissorhiza inaequalis*

incanus *in-KAN-nus*
incana, incanum
Grey, as in *Geranium incanum*

incarnatus *in-kar-NAH-tus*
incarnata, incarnatum
The colour of flesh, as in *Dactylorhiza incarnata*

Fritillaria imperialis,
crown imperial

incertus *in-KER-tus*
incerta, incertum
Doubtful; uncertain, as in *Draba incerta*

incisus *in-KYE-sus*
incisa, incisum
With deeply cut and irregular incisions, as in *Prunus incisa*

inclaudens *in-KLAW-denz*
Not closing, as in *Erepsia inclaudens*

inclinatus *in-klin-AH-tus*
inclinata, inclinatum
Bent downwards, as in *Moraea inclinata*

incomparabilis *in-kom-par-RAH-bih-lis*
incomparabilis, incomparabile
Incomparable, as in *Narcissus × incomparabilis*

incomptus *in-KOMP-tus*
incompta, incomptum
Without adornment, as in *Verbena incompta*

inconspicuus *in-kon-SPIK-yoo-us*
inconspicua, inconspicuum
Inconspicuous, as in *Hoya inconspicua*

incrassatus *in-kras-SAH-tus*
incrassata, incrassatum
Thickened, as in *Leucocoryne incrassata*

incurvatus *in-ker-VAH-tus*
incurvata, incurvatum

incurvus *in-ker-VUS*
incurva, incurvum
Bent inwards, as in *Carex incurva*

indicus *IN-dih-kus*
indica, indicum
Connected with India; may also apply to plants originating from the
East Indies or China, as in *Lagerstroemia indica*

indivisus *in-dee-VEE-sus*
indivisa, indivisum
Without divisions, as in *Cordyline indivisa*

induratus *in-doo-RAH-tus*
indurata, induratum
Hard, as in *Cotoneaster induratus*

inebrians *in-ee-BRI-enz*
Intoxicating, as in *Ribes inebrians*

inermis *IN-er-mis*
inermis, inerme
Without arms, e.g. without prickles, as in *Acaena inermis*

infaustus *in-FUS-tus*
infausta, infaustum
Unfortunate (sometimes therefore used of poisonous plants),
unlucky, as in *Colletia infausta*

infectorius *in-fek-TOR-ee-us*
infectoria, infectorium
Dyed; coloured, as in *Quercus infectoria*

infestus *in-FES-tus*
infesta, infestum
Dangerous; troublesome, as in *Melilotus infestus*

inflatus *in-FLAH-tus*
inflata, inflatum
Swollen up, as in *Codonopsis inflata*

infortunatus *in-for-tu-NAH-tus*
infortunata, infortunatum
Unfortunate (of poisonous plants), as in *Clerodendrum infortunatum*

infractus *in-FRAC-tus*
infracta, infractum
Curving inwards, as in *Masdevallia infracta*

infundibuliformis *in-fun-dih-bew-LEE-for-mis*
infundibuliformis, infundibuliforme
In the shape of a funnel or trumpet, as in
Crossandra infundibuliformis

infundibulus *in-fun-DIB-yoo-lus*
infundibula, infundibulum
A funnel, as in *Dendrobium infundibulum*

ingens *IN-genz*
Enormous, as in *Tulipa ingens*

inodorus *in-oh-DOR-us*
inodora, inodorum
Without scent, as in *Hypericum × inodorum*

inornatus *in-or-NAH-tus*
inornata, inornatum
Without ornament, as in *Boronia inornata*

*Masdevallia
infracta*

inquinans *in-KWIN-anz*
Polluted; stained; defiled, as in *Pelargonium inquinans*

insignis *in-SIG-nis*
insignis, insigne
Distinguished; remarkable, as in *Rhododendron insigne*

insititius *in-si-tih-TEE-us*
insititia, insititium
Grafted, as in *Prunus insititia*

insulanus *in-su-LAH-nus*
insulana, insulanum
insularis *in-soo-LAH-ris*
insularis, insulare
Relating to an island, as in *Tilia insularis*

integer *IN-teg-er*
integra, integrum
Entire, as in *Cyananthus integer*

integrifolius *in-teg-ree-FOH-lee-us*
integrifolia, integrifolium
With leaves that are complete, uncut, as in
Meconopsis integrifolia

intermedius *in-ter-MEE-dee-us*
intermedia, intermedium
Intermediate in colour, form or habit, as in *Forsythia × intermedia*

interruptus *in-ter-UP-tus*
interrupta, interruptum
Interrupted; not continuous, as in *Bromus interruptus*

intertextus *in-ter-TEKS-tus*
intertexta, intertextum
Intertwined, as in *Matucana intertexta*

intortus *in-TOR-tus*
intorta, intortum
Twisted, as in *Melocactus intortus*

intricatus *in-tree-KAH-tus*
intricata, intricatum
Tangled, as in *Asparagus intricatus*

intumescens *in-tu-MES-enz*
Swollen, as in *Carex intumescens*

intybaceus *in-tee-BAK-ee-us*
intybacea, intybaceum
Like chicory (*Cichorium intybus*), as in *Hieracium intybaceum*

inversus *in-VERS-us*
inversa, inversum
Turned over, as in *Quaqua inversa*

involucratus *in-vol-yoo-KRAH-tus*
involucrata, involucratum
With a ring of bracts surrounding several flowers, as in
Cyperus involucratus

Rubus idaeus,
raspberry

involutus *in-vol-YOO-tus*
involuta, involutum
Rolled inwards, as in *Gladiolus involutus*

ioensis *eye-oh-EN-sis*
ioensis, ioense
From Iowa, USA, as in *Malus ioensis*

ionanthus *eye-oh-NAN-thus*
ionantha, ionanthum
With violet-coloured flowers, as in *Saintpaulia ionantha*

ionopterus *eye-on-OP-ter-us*
ionoptera, ionopterum
With violet wings, as in *Koellensteinia ionoptera*

iridescens *ir-id-ES-enz*
Iridescent, as in *Phyllostachys iridescens*

iridiflorus *ir-id-uh-FLOR-us*
iridiflora, iridiflorum
With flowers like *Iris*, as in *Canna iridiflora*

iridifolius *ir-id-ih-FOH-lee-us*
iridifolia, iridifolium
With leaves like *Iris*, as in *Billbergia iridifolia*

iridioides *ir-id-ee-OY-deez*
Resembling *Iris*, as in *Dietes iridioides*

irregularis *ir-reg-yoo-LAH-ris*
irregularis, irregulare
With parts of different sizes, as in *Primula irregularis*

irriguus *ir-EE-gyoo-us*
irrigua, irriguum
Watered, as in *Pratia irrigua*

isophyllus *eye-so-FIL-us*
isophylla, isophyllum
With leaves of the same size, as in *Penstemon isophyllus*

italicus *ee-TAL-ih-kus*
italica, italicum
Connected with Italy, as in *Arum italicum*

ixioides *iks-ee-OY-deez*
Resembling corn lily (*Ixia*), as in *Libertia ixioides*

ixocarpus *iks-so-KAR-pus*
ixocarpa, ixocarpum
With sticky fruit, as in *Physalis ixocarpa*

J

jacobaeus *jak-koh-BAY-ee-us*
jacobaea, jacobaeum
Named for St James, or for Santiago (Cape Verde), as in
Senecio jacobaea

jackii *JAK-ee-eye*
Named after John George Jack (1861–1949), a Canadian
dendrologist at the Arnold Arboretum, Boston, USA, as in
Populus × jackii

jalapa *juh-LAP-a*
Connected with Xalapa, Mexico, as in *Mirabilis jalapa*

jamaicensis *ja-may-KEN-sis*
jamaicensis, jamaicense
From Jamaica, as in *Brunfelsia jamaicensis*

japonicus *juh-PON-ih-kus*
japonica, japonicum
Connected with Japan, as in *Cryptomeria japonica*

jasmineus *jaz-MIN-ee-us*
jasminea, jasmineum
Like jasmine (*Jasminum*), as in *Daphne jasminea*

jasminiflorus *jaz-min-IH-flor-us*
jasminiflora, jasminiflorum
With flowers like jasmine (*Jasminum*), as in *Rhododendron jasminiflorum*

jasminoides *jaz-min-OY-deez*
Resembling jasmine (*Jasminum*), as in *Trachelospermum jasminoides*

javanicus *juh-VAHN-ih-kus*
javanica, javanicum
Connected with Java, as in *Rhododendron javanicum*

jejunus *jeh-JOO-nus*
jejuna, jejunum
Small, as in *Eria jejuna*

jubatus *joo-BAH-tus*
jubata, jubatum
With awns, as in *Cortaderia jubata*

Mirabilis jalapa,
four o'clock plant

jucundus *joo-KUN-dus*
jucunda, jucundum
Agreeable; pleasing, as in *Osteospermum jucundum*

jugalis *joo-GAH-lis*
jugalis, jugale
jugosus *joo-GOH-sus*
jugosa, jugosum
Yoked, as in *Pahstia jugosa*

julaceus *joo-LA-see-us*
julacea, julaceum
With catkins, as in *Leucodon julaceus*

junceus *JUN-kee-us*
juncea, junceum
Like a rush, as in *Spartium junceum*

juniperifolius *joo-nip-er-ih-FOH-lee-us*
juniperifolia, juniperifolium
With leaves like a juniper (*Juniperus*), as in *Armeria juniperifolia*

juniperinus *joo-nip-er-EE-nus*
juniperina, juniperinum
Like a juniper (*Juniperus*); blue-black, as in *Grevillea juniperina*

juvenilis *joo-VEE-nil-is*
juvenilis, juvenile
Young, as in *Draba juvenilis*

JASMINUM

Few plants smell so sweetly or so strongly as *Jasminum*, commonly called simply jasmine or jessamine. In fact, scientific studies have proved that their perfume has a calming and soothing effect on animals as well as on people, although the belief that the smell also has aphrodisiac properties rather contradicts this. Belonging to the family *Oleaceae*, most jasmines are reasonably sized shrubs and climbers, although there are also low-mound-forming species such as *Jasminum parkeri* and *J. humile* f. *wallichianum*. Humilis, humilis, humile means low-growing or very small, while *wallichianus*, *wallichiana, wallichianum* commemorates the Danish botanist Nathaniel Wallich (1786–1854).

Although famed for their fragrant pure white flowers tinged with pink, other colours are also found in the genus. *J. beesianum* is a deciduous climber with lovely pinky red flowers that is quite big for the genus, growing up to 5 m (16½ ft); it is named after the English nursery known as Bees Ltd. As its name suggests, *J. polyanthum* is particularly free-flowering (*polyanthus, polyantha, polyanthum*, with many flowers), but it is not fully hardy so needs the protection of a sheltered site. One of the most useful jasmines is *J. nudiflorum*, the winter-flowering jasmine; its cheerful yellow flowers are borne on bare branches. *Nudiflorus, nudiflora, nudiflorum* refers to plants on

As its name implies, *Jasminum grandiflorum* has large flowers, which are white and extremely fragrant.

Jasminum angustifolium, wild jasmine

which the flowers appear before the leaves. Fully hardy, this is an undemanding plant with blooms that will brighten up even a cold and dark corner of the garden.

Plants with *jasmineus, jasminea, jasmineum* or *jasminoides* in their names will resemble *Jasminum* in some way, while the terms *jasminiflorus, jasminiflora* and *jasminiflorum* refer to their flowers. *Gardenia jasminoides* also goes by the names of common gardenia and cape jasmine. *Trachelospermum jasminoides* has a host of common names including star jasmine, Chinese jasmine and Confederate jasmine. *Solanum jasminoides* is the jasmine night-shade or, less poetically, the potato vine.

K

kamtschaticus *kam-SHAY-tih-kus*
kamtschatica, kamtschaticum
Connected with Kamchatka, Russia, as in *Sedum kamtschaticum*

kansuensis *kan-soo-EN-sis*
kansuensis, kansuense
From Kansu (Gansu), China, as in *Malus kansuensis*

karataviensis *kar-uh-taw-vee-EN-sis*
karataviensis, karataviense
From the Karatau mountains, Kazakhstan, as in
Allium karataviense

kashmirianus *kash-meer-ee-AH-nus*
kashmiriana, kashmirianum
Connected with Kashmir, as in *Actaea kashmiriana*

kermesinus *ker-mes-SEE-nus*
kermesina, kermesinum
Crimson, as in *Passiflora kermesina*

kewensis *kew-EN-sis*
kewensis, kewense
From Kew Gardens, London, England, as in *Primula kewensis*

Thunbergia kirkii,
blue sky shrub

Passiflora kermesina,
bitter-gourd-shaped passionfruit

kirkii *KIR-kee-eye*
Named after Thomas Kirk (1828–98), renowned botanist
of New Zealand flora; or after his son Harry Bower Kirk
(1903–1944), professor of biology at the University of New
Zealand; or after Sir John Kirk (1832–1922), botanist and
British Consul to Zanzibar, as in *Coprosma × kirkii* (after
Thomas Kirk)

kiusianus *key-oo-see-AH-nus*
kiusiana, kiusianum
Connected with Kyushu, Japan, as in
Rhododendron kiusianum

koreanus *kor-ee-AH-nus*
koreana, koreanum
Connected with Korea, as in *Abies koreana*

kurdicus *KUR-dih-kus*
kurdica, kurdicum
Referring to the Kurdish homeland in western Asia,
as in *Astragalus kurdicus*

L

labiatus *la-bee-AH-tus*
labiata, labiatum
Lipped, as in *Cattleya labiata*

labilis *LAH-bih-lis*
labilis, labile
Slippery; unstable, as in *Celtis labilis*

labiosus *lab-ee-OH-sus*
labiosa, labiosum
Lipped, as in *Besleria labiosa*

laburnifolius *luh-ber-nih-FOH-lee-us*
laburnifolia, laburnifolium
With leaves like *Laburnum*, as in *Crotalaria laburnifolia*

lacerus *LASS-er-us*
lacera, lacerum
Cut into fringe-like sections, as in *Costus lacerus*

laciniatus *la-sin-ee-AH-tus*
laciniata, laciniatum
Divided into narrow sections, as in *Rudbeckia laciniata*

lacrimans *LAK-ri-manz*
Weeping, as in *Eucalyptus lacrimans*

lacteus *lak-TEE-us*
Milk-white, as in *Cotoneaster lacteus*

lacticolor *lak-tee-KOL-or*
Milk-white, as in *Protea lacticolor*

lactiferus *lak-TIH-fer-us*
lactifera, lactiferum
Producing a milky sap, as in *Gymnema lactiferum*

lactiflorus *lak-tee-FLOR-us*
lactiflora, lactiflorum
With milk-white flowers, as in *Campanula lactiflora*

lacunosus *lah-koo-NOH-sus*
lacunosa, lacunosum
With deep holes or pits, as in *Allium lacunosum*

lacustris *lah-KUS-tris*
lacustris, lacustre
Relating to lakes, as in *Iris lacustris*

ladaniferus *lad-an-IH-fer-us*
ladanifera, ladaniferum
ladanifer *lad-an-EE-fer*
Producing ladanum, a fragrant gum resin, as in *Cistus ladanifer*

laetevirens *lay-tee-VY-renz*
Vivid green, as in *Parthenocissus laetevirens*

laetiflorus *lay-tee-FLOR-us*
laetiflora, laetiflorum
With bright flowers, as in *Helianthus × laetiflorus*

LATIN IN ACTION

A very hardy deciduous European tree or shrub that has wonderfully sweet-smelling flowers in May. The brightly coloured red berries that appear in such profusion in autumn are known as haws, hence the common name for the genus *Crataegus*, hawthorn. Like all *Crataegus*, *C. monogyna* is a very tough and hardy plant. Often referred to as quick or may, it makes an excellent stock-proof country hedge.

Crataegus laevigata,
hawthorn

laetus *LEE-tus*
laeta, laetum
Bright; vivid, as in *Pseudopanax laetum*

laevigatus *lee-vih-GAH-tus*
laevigata, laevigatum
laevis *LEE-vis*
laevis, laeve
Smooth, as in *Crocus laevigatus*

lagodechianus *la-go-chee-AH-nus*
lagodechiana, lagodechianum
Connected with Lagodekhi, Georgia, as in
Galanthus lagodechianus

lamellatus *la-mel-LAH-tus*
lamellata, lamellatum
Layered, as in *Vanda lamellata*

lanatus *la-NA-tus*
lanata, lanatum
Woolly, as in *Lavandula lanata*

lanceolatus *lan-see-oh-LAH-tus*
lanceolata, lanceolatum
lanceus *lan-SEE-us*
lancea, lanceum
In the shape of a spear, as in *Drimys lanceolata*

lanigerus *lan-EE-ger-rus*
lanigera, lanigerum
lanosus *LAN-oh-sus*
lanosa, lanosum
lanuginosus *lan-oo-gih-NOH-sus*
lanuginosa, lanuginosum
Woolly, as in *Leptospermum lanigerum*

lappa *LAP-ah*
A bur (prickly seed case or flower head), as in *Arctium lappa*

lapponicus *Lap-PON-ih-kus*
lapponica, lapponicum
lapponum *Lap-PON-num*
Connected with Lapland, as in *Salix lapponum*

laricifolius *lah-ris-ih-FOH-lee-us*
laricifolia, laricifolium
With leaves like a larch, as in *Penstemon laricifolius*

laricinus *lar-ih-SEE-nus*
laricina, laricinum
Like a larch (*Larix*), as in *Banksia laricina*

Caiophora lateritia

lasi-
Used in compound words to denote woolly

lasiandrus *las-ee-AN-drus*
lasiandra, lasiandrum
With woolly stamens, as in *Clematis lasiandra*

lasioglossus *las-ee-oh-GLOSS-us*
lasioglossa, lasioglossum
With a rough tongue, as in *Lycaste lasioglossa*

lateralis *lat-uh-RAH-lis*
lateralis, laterale
On the side, as in *Epidendrum laterale*

lateritius *la-ter-ee-TEE-us*
lateritia, lateritium
Brick-red, as in *Kalanchoe lateritia*

lati-
Used in compound words to denote broad

latiflorus *lat-ee-FLOR-us*
latiflora, latiflorum
With broad flowers, as in *Dendrocalamus latiflorus*

The Qualities of Plants

Beyond describing the purely physical properties of plants, such as size, colour, habit and fragrance, botanical Latin names also suggest other, often less tangible qualities that a plant may possess. Words abound that celebrate that most elusive of qualities: beauty. The aptly named beauty bush, *Kolkwitzia amabilis*, is one such (*amabilis, amabilis, amabile* meaning lovely). With its fragrant and exotic-looking flowers, *Gladiolus callianthus* lives up to its name (*callianthus, calliantha, callianthum*, with beautiful flowers). The term *callicarpa* tells us that a

Callicarpa dichotoma,
purple beautyberry

Also known as *Callicarpa purpurea*, the name refers to its beautiful purple fruits.

plant has beautiful fruit, as in the purple-fruited *Callicarpa bodinieri*, or beautyberry (*callicarpus, callicarpa, callicarpum*).

In the past when botanists wished to emphasise the impressive appearance of a plant, they would sometimes accord it a name that had aristocratic or royal associations. Thus a very distinguished plant would be given the epithet *imperialis* (*imperialis, imperiale*, very fine and showy), such as *Fritillaria imperialis*, commonly known as crown imperial or Kaiser's crown. Since ancient times, the sweet bay tree has been associated with emperors and heroic rulers, hence its name *Laurus nobilis*, meaning noble or renowned (*nobilis, nobilis, nobile*). *Agave victoriae-reginae* is named in honour of Queen Victoria, *victoriae-reginae* meaning of the queen Victoria. Other monarchical appellations include *basilicus* (*basilica, basilicum*) – princely or with royal properties; *rex*, relating to kings; *regalis* (*regalis, regale*), regal or of exceptional merit and *regius* (*regia, regium*), simply meaning royal.

It was not unknown for botanists in the past to wander into the realms of sentiment; some apportioned almost human qualities to the plants they named. Often it is far from clear why such names were chosen, as with *Cassia fastuosa* – the species name meaning proud (*fastuosus, fastuosa, fastuosum*). Similarly, it may seem strange that the night-scented pelargonium is called *Pelargonium triste* (*tristis, tristis, triste*, meaning dull or sad), until we remember that compared to its showy relatives it does indeed have dullish flowers. Sometimes an epithet misleads by its similarity to an English word. Thus the name *Sorbus vexans* (a kind of whitebeam) could seem to imply an annoying quality, whereas it simply means the tree's identity was a vexed question for many years. Turning to poetry, classical literature is the source for the pheasant's eye daffodil, *Narcissus poeticus*, named after the

Senecio elegans,
purple ragwort

Elegans means elegant and usually refers to a plant's flowers,
as in the purple blooms of this annual.

Adonis vernalis,
spring pheasant's eye

Vernalis means of the spring; in this case it refers to the
yellow flowers that appear in spring.

beautiful but vain Narcissus of Greek mythology, who
the gods punished by turning into a flower (*poeticus,
poetica, poeticum,* of poets).

On a more practical note, the epithets *futilis* and
inutilis indicate that a plant has no use (*futilis, futilis,
futile* and *inutilis, inutilis, inutile*). By contrast, *utilis*
and *utilissimus* both suggest that a plant has some
culinary, medicinal or economic use or value, as with
the leaves of *Pandanus utilis,* the common screw pine,
which are harvested and used to make mats. (*Utilis,
utilis, utile* and *utilissimus, utilissima, utilissimum.*) A
plant's name may denote its produce, such as *sac-
chiferus* (*sacchifera, sacchiferum*), producing sugar,
gummifer (*gummifera, gummiferum*), producing gum
or resin, and *viniferus* (*vinifera, viniferum*), producing
wine, as in *Vitis vinifera,* the grape vine.

Several descriptive names relate to lifespan, time or
the seasons. *Monocarpus* and *hapaxanthus* plants bear
fruit once then die (*monocarpus, monocarpa, monocar-
pum*). *Diurnus* (*diurna, diurnum*) means flowering by
day, while *noctiflorus* (*noctiflora, noctiflorum*) means
flowering at night. *Aestivalis* relates to summer and
hibernalis to winter (*aestivalis, aestivalis, aestivale* and
hibernalis, hibernalis, hibernale). *Hibernus* (*hiberna,
hibernum*) denotes that a plant is winter-green or
winter-flowering and *hiemalis* (*hiemalis, hiemale*)
means of the winter, like *Camellia hiemalis,* which
blooms in the winter months. Likewise, the winter
aconite *Eranthis hyemalis* bears cheerful yellow
flowers, making them a very welcome sight in the dark
days of late winter (*hyemalis, hyemalis, hyemale,*
relating to winter).

latifolius *lat-ee-FOH-lee-us*
latifolia, latifolium
With broad leaves, as in *Lathyrus latifolius*

latifrons *lat-ee-FRONS*
With broad fronds, as in *Encephalartos latifrons*

latilobus *lat-ee-LOH-bus*
latiloba, latilobum
With wide lobes, as in *Campanula latiloba*

latispinus *la-tih-SPEE-nus*
latispina, latispinum
With wide thorns, as in *Ferocactus latispinus*

LATIN IN ACTION

With its simple saucer-shaped flowers of purple, violet-blue or white, this herbaceous perennial is a perfect plant for cottage gardens. It can reach a stately height of 1 m (3 ft). To preserve the intensity of the flower colour, it is best to avoid planting it in full sun.

Campanula latiloba,
great bellflower

---

laudatus *law-DAH-tus*
laudata, laudatum
Worthy of praise. as in *Rubus laudatus*

laurifolius *law-ree-FOH-lee-us*
laurifolia, laurifolium
With leaves like bay (*Laurus*), as in *Cistus laurifolius*

laurinus *law-REE-nus*
laurina, laurinum
Like a laurel or bay tree (*Laurus*), as in *Hakea laurina*

laurocerasus *law-roh-KER-uh-sus*
From the Latin for cherry and laurel, as in *Prunus laurocerasus*

lavandulaceus *la-van-dew-LAY-see-us*
lavandulacea, lavandulaceum
Like lavender (*Lavandula*), as in *Chirita lavandulacea*

lavandulifolius *lav-an-dew-lih-FOH-lee-us*
lavandulifolia, lavandulifolium
With leaves like lavender (*Lavandula*), as in *Salvia lavandulifolia*

laxiflorus *laks-ih-FLO-rus*
laxiflora, laxiflorum
With loose, open flowers, as in *Lobelia laxiflora*

laxifolius *laks-ih-FOH-lee-us*
laxifolia, laxifolium
With loose, open leaves, as in *Athrotaxis laxifolia*

laxus *LAX-us*
laxa, laxum
Loose; open, as in *Freesia laxa*

ledifolius *lee-di-FOH-lee-us*
ledifolia, ledifolium
With leaves like *Ledum*, as in *Ozothamnus ledifolius*

leianthus *lee-AN-thus*
leiantha, leianthum
With smooth flowers, as in *Bouvardia leiantha*

leiocarpus *lee-oh-KAR-pus*
leiocarpa, leiocarpum
With smooth fruit, as in *Cytisus leiocarpus*

leiophyllus *lay-oh-FIL-us*
leiophylla, leiophyllum
With smooth leaves, as in *Pinus leiophylla*

leichtlinii *leekt-LIN-ee-eye*
Named after Max Leichtlin (1831–1910), German plant collector
from Baden-Baden, Germany, as in *Camassia leichtlinii*

lentiginosus *len-tig-ih-NOH-sus*
lentiginosa, lentiginosum
With freckles, as in *Coelogyne lentiginosa*

lentus *LEN-tus*
lenta, lentum
Tough but flexible, as in *Betula lenta*

leonis *le-ON-is*
With the colour of, or teeth like, a lion, as in *Angraecum leonis*

leontoglossus *le-on-toh-GLOSS-us*
leontoglossa, leontoglossum
With a throat or tongue of a lion, as in *Masdevallia leontoglossa*

leonurus *lee-ON-or-us*
leonura, leonurum
Like a lion's tail, as in *Leonotis leonurus*

leopardinus *leh-par-DEE-nus*
leopardina, leopardinum
With spots like a leopard, as in *Calathea leopardina*

lepidus *le-PID-us*
lepida, lepidum
Graceful; elegant, as in *Lupinus lepidus*

lept-
Used in compound words to denote thin or slender

leptanthus *lep-TAN-thus*
leptantha, leptanthum
With slender flowers, as in *Colchicum leptanthum*

leptocaulis *lep-toh-KAW-lis*
leptocaulis, leptocaule
With thin stems, as in *Cylindropuntia leptocaulis*

leptocladus *lep-toh-KLAD-us*
leptoclada, leptocladum
With thin branches, as in *Acacia leptoclada*

leptophyllus *lep-toh-FIL-us*
leptophylla, leptophyllum
With thin leaves, as in *Cassinia leptophylla*

Lathyrus latifolius,
broad-leaved everlasting pea

leptosepalus *lep-toh-SEP-a-lus*
leptosepala, leptosepalum
With thin sepals, as in *Caltha leptosepala*

leptopus *LEP-toh-pus*
With thin stalks, as in *Antigonon leptopus*

leptostachys *lep-toh-STAH-kus*
With slender spikes, as in *Aponogeton leptostachys*

leuc-
Used in compound words to denote white

leucanthus *lew-KAN-thus*
leucantha, leucanthum
With white flowers, as in *Tulbaghia leucantha*

leucocephalus *loo-koh-SEF-uh-lus*
leucocephala, leucocephalum
With a white head, as in *Leucaena leucocephala*

leucochilus *loo-KOH-ky-lus*
leucochila, leucochilum
With white lips, as in *Oncidium leucochila*

leucodermis *loo-koh-DER-mis*
leucodermis, leucoderme
With white skin, as in *Pinus leucodermis*

leuconeurus *loo-koh-NOOR-us*
leuconeura, leuconeurum
With white nerves, as in *Maranta leuconeura*

leucophaeus *loo-koh-FAY-us*
leucophaea, leucophaeum
Dusky white, as in *Dianthus leucophaeus*

leucophyllus *loo-koh-FIL-us*
leucophylla, leucophyllum
With white leaves, as in *Sarracenia leucophylla*

leucorhizus *loo-koh-RYE-zus*
leucorhiza, leucorhizum
With white roots, as in *Curcuma leucorhiza*

leucoxanthus *loo-koh-ZAN-thus*
leucoxantha, leucoxanthum
Whitish yellow, as in *Sobralia leucoxantha*

leucoxylon *loo-koh-ZY-lon*
With white wood, as in *Eucalyptus leucoxylon*

libani *LIB-an-ee*
libanoticus *lib-an-OT-ih-kus*
libanotica, libanoticum
Connected with Mount Lebanon, Lebanon, as in *Cedrus libani*

libericus *li-BEER-ih-kus*
liberica, libericum
From Liberia, as in *Coffea liberica*

liburnicus *li-BER-nih-kus*
liburnica, liburnicum
Connected with Liburnia (now in Croatia), as in
Asphodeline liburnica

lignosus *lig-NOH-sus*
lignosa, lignosum
Woody, as in *Tuberaria lignosa*

ligularis *lig-yoo-LAH-ris*
ligularis, ligulare
ligulatus *lig-yoo-LAIR-tus*
ligulata, ligulatum
Shaped like a strap, as in *Acacia ligulata*

ligusticifolius *lig-us-tih-kih-FOH-lee-us*
ligusticifolia, ligusticifolium
With leaves like lovage, as in *Clematis ligusticifolia*

ligusticus *lig-US-tih-kus*
ligustica, ligusticum
Connected with Liguria, Italy, as in *Crocus ligusticus*

ligustrifolius *lig-us-trih-FOH-lee-us*
ligustrifolia, ligustrifolium
With leaves like privet (*Ligustrum*), as in *Hebe ligustrifolia*

ligustrinus *lig-us-TREE-nus*
ligustrina, ligustrinum
Like privet (*Ligustrum*), as in *Ageratina ligustrina*

lilacinus *ly-luc-SEE-nus*
lilacina, lilacinum
Lilac, as in *Primula lilacina*

lili-
Used in compound words to denote lily

liliaceus *lil-lee-AY-see-us*
liliacea, liliaceum
Like lily (*Lilium*), as in *Fritillaria liliacea*

liliiflorus *lil-lee-ih-FLOR-us*
liliiflora, liliiflorum
With flowers like lily (*Lilium*), as in *Magnolia liliiflora*

liliifolius *lil-ee-eye-FOH-lee-us*
liliifolia, liliifolium
With leaves like lily (*Lilium*), as in *Adenophora liliifolia*

limbatus *lim-BAH-tus*
limbata, limbatum
Bordered, as in *Primula limbata*

limensis *lee-MEN-sis*
limensis, limense
From Lima, Peru, as in *Haageocereus limensis*

limoniifolius *lim-on-ih-FOH-lee-us*
limoniifolia, limoniifolium
With leaves like sea lavender (*Limonium*), as in *Asyneuma limoniifolium*

limosus *lim-OH-sus*
limosa, limosum
Growing in marshy or muddy habitats, as in *Carex limosa*

linariifolius *lin-ar-ee-FOH-lee-us*
linariifolia, linariifolium
With leaves like toadflax (*Linaria*), as in *Melaleuca linariifolia*

lindleyanus *lind-lee-AH-nus*
lindleyana, lindleyanum
lindleyi *lind-lee-EYE*
Named after John Lindley (1799–1865), English botanist associated
with the Royal Horticultural Society, as in *Buddleja lindleyana*

linearis *lin-AH-ris*
linearis, lineare
With narrow, almost parallel sides, as in *Ceropegia linearis*

lineatus *lin-ee-AH-tus*
lineata, lineatum
With lines or stripes, as in *Rubus lineatus*

lingua *LIN-gwa*
A tongue or like a tongue, as in *Pyrrosia lingua*

linguiformis *lin-gwih-FORM-is*
linguiformis, linguiforme
lingulatus *lin-gyoo-LAH-tus*
lingulata, lingulatum
Shaped like a tongue, as in *Guzmania lingulata*

liniflorus *lin-ih-FLOR-us*
liniflora, liniflorum
With flowers like flax (*Linum*), as in *Byblis liniflora*

linifolius *lin-ih-FOH-lee-us*
linifolia, linifolium
With leaves like flax (*Linum*), as in *Tulipa linifolia*

linnaeanus *lin-ee-AH-nus*
linnaeana, linnaeanum
linnaei *lin-ee-eye*
Named after Carl Linnaeus (1707–78), Swedish botanist,
as in *Solanum linnaeanum*

linoides *li-NOY-deez*
Resembling flax (*Linum*), as in *Monardella linoides*

litangensis *lit-ang-EN-sis*
litangensis, litangense
From Litang, China, as in *Lonicera litangensis*

lithophilus *lith-oh-FIL-us*
lithophila, lithophilum
Growing in rocky habitats, as in *Anemone lithophila*

littoralis *lit-tor-AH-lis*
littoralis, littorale
littoreus *lit-TOR-ee-us*
littorea, littoreum
Growing by the sea, as in *Griselinia littoralis*

The genus of trees and shrubs known as *Hakea* is
named after the German patron of botany, Baron
Christian Ludwig von Hake (1745–1818). Native
to the sandy swamps and wetlands of southwestern
Australia, they are now in wide cultivation across the
country. *Hakea linearis* has creamy white flowers;
the epithet *linearis*, meaning narrow, refers to the
shape of the leaves. Likewise, the leaves of *Aspalathus
linearis* and *Chilopsis linearis* are similarly long and
narrow. *A. linearis* is the rooibos tree from the western
Cape of South Africa. *C. linearis* is commonly known
as the desert willow and is found in the southwestern
United States and Mexico. *Flaveria linearis*, known
as yellowtop, is a native of Florida; it too has long,
narrow leaves and bright yellow flowers, which are a
particularly rich source of nectar for insects.

Hakea linearis

lividus *LI-vid-us*
livida, lividum
Blue-grey; the colour of lead, as in *Helleborus lividus*

lobatus *low-BAH-tus*
lobata, lobatum
With lobes, as in *Cyananthus lobatus*

lobelioides *lo-bell-ee-OH-id-ees*
Resembling *Lobelia*, as in *Wahlenbergia lobelioides*

LATIN IN ACTION

In Victorian Britain, a frenzied craze for collecting ferns broke out among a large number of gardeners. Known as *pteridomania*, or fern-fever, the love for all things fern-like spread to the decorative arts, and images of ferns appeared on ceramics, textiles and even on iron garden benches, the deep lobes, or divisions, of the leaves creating wonderful shapes.

Polystichum aculeatum (syn. *P. lobatum*), hard shield fern

---

lobophyllus *lo-bo-FIL-us*
lobophylla, lobophyllum
With lobed leaves, as in *Viburnum lobophyllum*

lobularis *lobe-yoo-LAY-ris*
lobularis, lobulare
With lobes, as in *Narcissus lobularis*

lobulatus *lob-yoo-LAH-tus*
lobulata, lobulatum
With small lobes, as in *Crataegus lobulata*

loliaceus *loh-lee-uh-SEE-us*
loliacea, loliaceum
Like rye-grass (*Lolium*), as in × *Festulolium loliaceum*

longibracteatus *lon-jee-brak-tee-AH-tus*
longibracteata, longibracteatum
With long bracts, as in *Pachystachys longibracteata*

longipedunculatus *long-ee-ped-un-kew-LAH-tus*
longipedunculata, longipedunculatum
With a long peduncle, as in *Magnolia longipedunculata*

longicaulis *lon-jee-KAW-lis*
longicaulis, longicaule
With long stalks, as in *Aeschynanthus longicaulis*

longicuspis *lon-jih-kus-pis*
longicuspis, longicuspe
With long points, as in *Rosa longicuspis*

longiflorus *lon-jee-FLO-rus*
longiflora, longiflorum
With long flowers, as in *Crocus longiflorus*

longifolius *lon-jee-FOH-lee-us*
longifolia, longifolium
With long leaves, as in *Pulmonaria longifolia*

longilobus *lon-JEE-loh-bus*
longiloba, longilobum
With long lobes, as in *Alocasia longiloba*

longipes *LON-juh-peez*
With a long stalk, as in *Acer longipes*

longipetalus *lon-jee-PET-uh-lus*
longipetala, longipetalum
With long petals, as in *Matthiola longipetala*

longiracemosus *lon-jee-ray-see-MOH-sus*
longiracemosa, longiracemosum
With long racemes, as in *Incarvillea longiracemosa*

longiscapus *lon-jee-SKAY-pus*
longiscapa, longiscapum
With a long scape, as in *Vriesea longiscapa*

longisepalus *lon-jee-SEE-pal-us*
longisepala, longisepalum
With long sepals, as in *Allium longisepalum*

longispathus *lon-jis-PAY-thus*
longispatha, longispathum
With long spathes, as in *Narcissus longispathus*

longissimus *lon-JIS-ih-mus*
longissima, longissimum
Very long, as in *Aquilegia longissima*

longistylus *lon-jee-STY-lus*
longistyla, longistylum
With a long style, as in *Arenaria longistyla*

longus *LONG-us*
longa, longum
Long, as in *Cyperus longus*

lophanthus *low-FAN-thus*
lophantha, lophanthum
With crested flowers, as in *Paraserianthes lophantha*

lotifolius *lo-tif-FOH-lee-us*
lotifolia, lotifolium
With leaves like *Lotus*, as in *Goodia lotifolia*

louisianus *loo-ee-see-AH-nus*
louisiana, louisianum
Connected with Louisiana, USA, as in *Proboscidea louisiana*

lucens *LOO-senz*
lucidus *LOO-sid-us*
lucida, lucidum
Bright; shining; clear, as in *Ligustrum lucidum*

ludovicianus *loo-doh-vik-ee-AH-nus*
ludoviciana, ludovicianum
Connected with Louisiana, USA, as in
Artemisia ludoviciana

lunatus *loo-NAH-tus*
lunata, lunatum
lunulatus *loo-nu-LAH-tus*
lunulata, lunulatum
Shaped like the crescent moon, as in *Cyathea lunulata*

lupulinus *lup-oo-LEE-nus*
lupulina, lupulinum
Like hop (*Humulus lupulus*), as in *Medicago lupulina*

luridus *LEW-rid-us*
lurida, luridum
Pale yellow, wan, as in *Moraea lurida*

lusitanicus *loo-si-TAN-ih-kus*
lusitanica, lusitanicum
Connected with Lusitania (Portugal and some parts of Spain), as in
Prunus lusitanica

luteolus *loo-tee-OH-lus*
luteola, luteolum
Yellowish, as in *Primula luteola*

lutetianus *loo-tee-shee-AH-nus*
lutetiana, lutetianum
Connected with Lutetia (Paris), France, as in
Circaea lutetiana

luteus *LOO-tee-us*
lutea, luteum
Yellow, as in *Calochortus luteus*

luxurians *luks-YOO-ee-anz*
Luxuriant, as in *Begonia luxurians*

lycius *LY-cee-us*
lycia, lycium
Connected with Lycia, now part of Turkey, as in *Phlomis lycia*

lycopodioides *ly-kop-oh-dee-OY-deez*
Resembling clubmoss (*Lycopodium*), as in
Cassiope lycopodioides

lydius *LID-ee-us*
lydia, lydium
Connected with Lydia, now part of Turkey, as in *Genista lydia*

lysimachioides *ly-see-mak-ee-OY-deez*
Resembling loosestrife (*Lysimachia*), as in *Hypericum lysimachioides*

LYCOPERSICON

There is a huge range of colour, shape and size of tomatoes for the home grower to choose from.

The large number of plants whose botanical names are connected with animals often surprises gardeners. For instance, the Greek word *alopekouros* means fox's tail and has given us the Latin name for *Alopecurus*, the foxtail grass or meadow foxtail. The species epithet for horse chestnut (*Aesculus hippocastanum*), on the other hand, derives from Greek *hippos* (horse) and *kastanos* (chestnut), and probably relates to the former use of its seeds (conkers) in the treatment of equine respiratory problems. Perhaps most surprising is the derivation of the species name for tomato, *lycopersicum*. Somewhat unexpectedly, the Greek *lykos* means wolf and *persicon* peach so it is literally a wolfpeach!

The ancient Aztecs and Incas happily ate wild forms of tomato and these were also consumed by Native Americans. However, the plant was greeted with much suspicion when it was introduced into Europe from South America by returning Spanish conquistadores, and many thought it (along with the potato) was poisonous. *Solanum lycopersicum* belongs to the *Solanaceae* family, which also contains potatoes, peppers, tobacco and deadly nightshade.

It is possibly this association that led people to at first treat the plant as primarily ornamental rather than edible. Other plant names with the prefix *lyco-* include *Lycopodium* and *Lycopus* (both meaning wolf's foot in Greek), while the common name for the highly toxic *Aconitum lycoctonum* is wolfsbane (Greek *ktonos*, murder). Wolves and poisons apart, the tomato, though deemed a vegetable in an American Supreme Court ruling of 1893, is botanically a fruit. Certainly it has linguistic links to other fruits; the size and shape of various forms are known as cherry and plum and perhaps its loveliest common name is love apple.

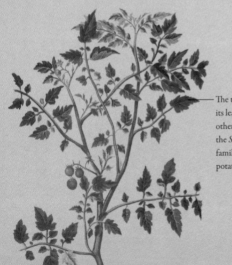

The tomato shares its leaf shape with other members of the *Solanaceae* family such as the potato.

M

macedonicus *mas-eh-DON-ih-kus*
macedonica, macedonicum
Connected with Macedonia, as in *Knautia macedonica*

macilentus *mas-il-LEN-tus*
macilenta, macilentum
Thin; lean, as in *Justicia macilenta*

macro-
Used in compound words to denote either long or large

macracanthus *mak-ra-KAN-thus*
macracantha, macracanthum
With large spines, as in *Acacia macracantha*

macrandrus *mak-RAN-drus*
macrandra, macrandrum
With large anthers, as in *Eucalyptus macrandra*

macranthus *mak-RAN-thus*
macrantha, macranthum
With large flowers, as in *Hebe macrantha*

macrobotrys *mak-ro-BOT-rees*
With large grape-like clusters, as in *Strongylodon macrobotrys*

macrocarpus *ma-kro-KAR-pus*
macrocarpa, macrocarpum
With large fruit, as in *Cupressus macrocarpa*

macrocephalus *mak-roh-SEF-uh-lus*
macrocephala, macrocephalum
With large heads, as in *Centaurea macrocephala*

macrodontus *mak-roh-DON-tus*
macrodonta, macrodontum
With large teeth, as in *Olearia macrodonta*

macromeris *mak-roh-MER-is*
With many or large parts, as in *Coryphantha macromeris*

macrophyllus *mak-roh-FIL-us*
macrophylla, macrophyllum
With large or long leaves, as in *Hydrangea macrophylla*

macropodus *mak-roh-POH-dus*
macropoda, macropodum
With stout stalks, as in *Daphniphyllum macropodum*

macrorrhizus *mak-roh-RY-zus*
macrorrhiza, macrorrhizum
With large roots, as in *Geranium macrorrhizum*

macrospermus *mak-roh-SPERM-us*
macrosperma, macrospermum
With large seeds, as in *Senecio macrospermus*

macrostachyus *mak-ro-STAH-kus*
macrostachya, macrostachyum
With long or large spikes, as in *Setaria macrostachya*

maculatus *mak-yuh-LAH-tus*
maculata, maculatum
maculosus *mak-yuh-LAH-sus*
maculosa, maculosum
With spots, as in *Begonia maculata*

Arum maculatum,
lords and ladies

madagascariensis *mad-uh-gas-KAR-ee-EN-sis*
madagascariensis, madagascariense
From Madagascar, as in *Buddleja madagascariensis*

maderensis *ma-der-EN-sis*
maderensis, maderense
From Madeira, as in *Geranium maderense*

magellanicus *ma-jell-AN-ih-kus*
megallanica, megallanicum
Connected with the Straits of Magellan, South America,
as in *Fuchsia magellanica*

LATIN IN ACTION

Convallaria majalis is also known as Our Lady's tears
or Mary's tears in some regions, possibly due to the
pure white tear-shaped flowers. *Majalis* tells us that it
flowers in May; traditionally, the flowers are sold on
the streets of France on the first day of May.

Convallaria majalis,
lily-of-the-valley

---

magellensis *mag-ah-LEN-sis*
magellensis, magellense
Of the Maiella massif, Italy, as in *Sedum magellense*

magnificus *mag-NIH-fih-kus*
magnifica, magnificum
Splendid, magnificent, as in *Geranium* × *magnificum*

magnus *MAG-nus*
magna, magnum
Great; big, as in *Alberta magna*

majalis *maj-AH-lis*
majalis, majale
Flowering in May, as in *Convallaria majalis*

major *MAY-jor*
major, majus
Bigger; larger, as in *Astrantia major*

malabaricus *mal-uh-BAR-ih-kus*
malabarica, malabaricum
Connected with the Malabar coast, India, as in
Bauhinia malabarica

malacoides *mal-a-koy-deez*
Soft, as in *Erodium malacoides*

malacospermus *mal-uh-ko-SPER-mus*
malacosperma, malacospermum
With soft seeds, as in *Hibiscus malacospermus*

maliformis *ma-lee-for-mees*
maliformis, maliforme
Shaped like an apple, as in *Passiflora maliformis*

malvaceus *mal-VAY-see-us*
malvacea, malvaceum
Like mallow (*Malva*), as in *Physocarpus malvaceus*

malviflorus *mal-VEE-flor-us*
malviflora, malviflorum
With flowers like mallow (*Malva*), as in *Geranium malviflorum*

malvinus *mal-VY-nus*
malvina, malvinum
Mauve, as in *Plectranthus malvinus*

mammillatus *mam-mil-LAIR-tus*
mammillata, mammillatum
mammillaris *mam-mil-LAH-ris*
mammillaris, mammillare
mammosus *mam-OH-sus*
mammosa, mammosum
Bearing nipple- or breast-like structures, as in *Solanum mammosum*

mandshuricus *mand-SHEU-rih-kus*
mandshurica, mandshuricum
manshuricus *man-SHEU-rih-kus*
manshurica, manshuricum
Connected with Manchuria, northeast Asia, as in *Tilia mandshurica*

manicatus *mah-nuh-KAH-tus*
manicata, manicatum
With long sleeves, as in *Gunnera manicata*

margaritaceus *mar-gar-ee-tuh-KEE-us*
margaritacea, margaritaceum
margaritus *mar-gar-ee-tus*
margarita, margaritum
Relating to pearls, as in *Anaphalis margaritacea*

margaritiferus *mar-guh-rih-TIH-fer-us*
margaritifera, margaritiferum
With pearls, as in *Haworthia margaritifera*

marginalis *mar-gin-AH-lis*
marginalis, marginale
marginatus *mar-gin-AH-tus*
marginata, marginatum
Margined, as in *Saxifraga marginata*

marianus *mar-ee-AH-nus*
mariana, marianum
Of the Virgin Mary (or sometimes Maryland, USA), as in
Silybum marianum

marilandicus *mar-i-LAND-ih-kus*
marilandica, marilandicum
Connected with Maryland, USA, as in *Quercus marilandica*

maritimus *muh-RIT-tim-mus*
maritima, maritimum
Relating to the sea, as in *Armeria maritima*

marmoratus *mar-mor-RAH-tus*
marmorata, marmoratum
marmoreus *mar-MOH-ree-us*
marmorea, marmoreum
Marbled; mottled, as in *Kalanchoe marmorata*

Spigelia marilandica,
Indian pink

maroccanus *mar-oh-KAH-nus*
maroccana, maroccanum
Connected with Morocco, as in *Linaria maroccana*

martagon *MART-uh-gon*
A word of uncertain origin, thought in *Lilium martagon* to refer to
the turban-like flower

martinicensis *mar-teen-i-SEN-sis*
martinicensis, martinicense
From Martinique, Lesser Antilles, as in *Trimezia martinicensis*

mas *MAS*
masculus *MASK-yoo-lus*
mascula, masculum
With masculine qualities, male, as in *Cornus mas*

matronalis *mah-tro-NAH-lis*
matronalis, matronale
Relating to 1st March, the Roman Matronalia, festival of motherhood
and Juno as goddess of childbirth, as in *Hesperis matronalis*

mauritanicus *maw-rih-TAWN-ih-kus*
mauritanica, mauritanicum
Connected with North Africa, especially Morocco,
as in *Lavatera mauritanica*

CARL LINNAEUS

(1707–78)

If gardeners ever despair at the prospect of remembering the Latin name of a plant, they should pause and give thanks to the 18th-century botanist, physician and zoologist Carl Linnaeus. It is due to his rigorous rationalisation of plant names that they now need only recall two words, rather than the dozen or so that were commonly used in previous eras.

Born in the southern province of Småland, Sweden, the son of a country parson, Linnaeus was raised in a family in which Latin was spoken daily. After showing an early interest in plants and botany, he studied medicine at Uppsala University, a subject closely allied to herbalism at that time. Linnaeus possessed an insatiable curiosity about all aspects of the world around him. After classifying the plants, animals and minerals of his native country, he travelled widely, including trips to England, Holland and Lapland. Fortunately his highly detailed and illustrated notebooks survive from the 4,600-mile-long journey he took across Lapland. They show his keen observation of the native flora and fauna he encountered on the trip, including the hundred or so new botanical species he discovered. Returning to Uppsala as professor of botany, he was regarded as an inspiring teacher. Many of his students, known as Apostles, went on to make important scientific discoveries all over the world. Today Linnaeus is best remembered for the binomial,

Among his contemporaries, Linnaeus was famed for his insatiable curiosity about the natural world and for the accuracy of his visual memory.

or two-word, system of naming plants that he developed and refined from the earlier work of several 17th-century scientists, most notably Caspar Bauhin (1560–1624). Thanks to Linnaeus and his predecessors, plants, and indeed animals, are now classified into categories of kingdom, class, order, family, genus and species. Of most use to gardeners is the binomial method of nomenclature, whereby a plant is first attributed to a particular genus then given its specific species name. The species may then be subdivided into subspecies, variety and form for greater clarity of identification. Occasionally one may also see a person's name included, as for instance in *Pelargonium zonale* (Linnaeus). This tells us that Linnaeus was the first person to describe that particular plant, though not necessarily to discover it.

Linnaeus based his classification of plants on their sexual characteristics, dividing plants into groups depending on their number of stamens and pistils (the male and female sexual organs of plants). He was aware that this was an artificial structure and after his death it was superseded by a natural botanical system. This emphasis on the reproductive aspects of plants led Linnaeus to use some rather fanciful language, describing the world of plants in terms of 'brides', 'bridegrooms' and 'bridal beds'.

Linnaeus published numerous works throughout his career. Among the most influential is *Systema Naturae* (1735); this was originally produced as a pamphlet outlining his new system of classifying the natural world. He continued to extend the work over the following decades, until it became a two-volume publication in 1758. His *Genera Plantarum* (1737) describes in detail all the 935 plant genera that were then known. This was followed in 1753 by *Species Plantarum*; describing thousands of plant species, it became the basis for modern nomenclature. Linnaeus's system of classification enabled scientists to fit previously unidentified plants and animals into a sound framework of knowledge, based on empirical observation. Thus they could begin to see how one species related to another. This was of particular importance at the time, as it was just at the moment when huge quantities of new and diverse plant specimens were being introduced into Europe from all over the world.

The importance of Linnaeus's work was recognised in his lifetime. He became Court Physician in 1747, was made a Knight of the Polar Star in 1758, then ennobled in 1761, taking the title of Carl von Linné. After a series of debilitating strokes, he died aged 71. Methodical, practical and rational, Linnaeus was, despite the occasional literary flight of fancy, a master of precise and accurate simplification.

Linnaea borealis, twinflower

Linnaea borealis (*borealis, borealis, boreale,* meaning northern) is one of the few plants named after Carl Linnaeus. It was one of his favourite plants and has pretty bell-shaped flowers that hang in pairs from a single stem. Commonly known as the twinflower, it is at home growing in forest habitats; its appearance is often an indicator of ancient woodland.

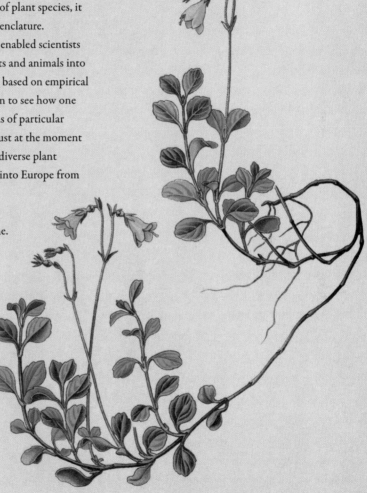

'LINNAEUS WAS IN REALITY A POET WHO HAPPENED TO BECOME A NATURALIST.'
August Strindberg (1849–1912)

mauritianus *maw-rih-tee-AH-nus*
mauritiana, mauritianum
Connected with Mauritius, Indian Ocean, as in *Croton mauritianus*

maxillaris *max-ILL-ah-ris*
maxillaris, maxillare
Relating to the jaw, as in *Zygopetalum maxillare*

maximus *MAKS-ih-mus*
maxima, maximum
Largest, as in *Rudbeckia maxima*

medicus *MED-ih-kus*
medica, medicum
Medicinal, as in *Citrus medica*

mediopictus *MED-ee-o-pic-tus*
mediopicta, mediopictum
With a stripe or colour running down the middle, as in
Calathea mediopicta

mediterraneus *med-e-ter-RAY-nee-us*
mediterranea, mediterraneum
Either from a land-locked region, or from the Mediterranean, as in
Minuartia mediterranea

medius *MEED-ee-us*
media, medium
Intermediate; middle, as in *Mahonia × media*

medullaris *med-yoo-LAH-ris*
medullaris, medullare
medullus *med-DUL-us*
medulla, medullum
Pithy, as in *Cyathea medullaris*

mega-
Used in compound words to denote big

megacanthus *meg-uh-KAN-thus*
megacantha, megacanthum
With big spines, as in *Opuntia megacantha*

megacarpus *meg-uh-CAR-pus*
megacarpa, megacarpum
With big fruits, as in *Ceanothus megacarpus*

megalanthus *meg-uh-LAN-thus*
megalantha, megalanthum
With big flowers, as in *Potentilla megalantha*

megalophyllus *meg-uh-luh-FIL-us*
megalophylla, megalophyllum
With big leaves, as in *Ampelopsis megalophylla*

megapotamicus *meg-uh-poh-TAM-ih-kus*
megapotamica, megapotamicum
Connected with a big river, for instance the Amazon or Rio Grande,
as in *Abutilon megapotamicum*

megaspermus *meg-uh-SPER-mus*
megasperma, megaspermum
With big seeds, as in *Callerya megasperma*

megastigma *meg-a-STIG-ma*
With a big stigma, as in *Boronia megastigma*

melanocaulon *mel-an-oh-KAW-lon*
With black stems, as in *Blechnum melanocaulon*

melanocentrus *mel-an-oh-KEN-trus*
melanocentra, melanocentrum
With a black centre, as in *Saxifraga melanocentra*

melanococcus *mel-an-oh-KOK-us*
melanococca, melanococcum
With black berries, as in *Elaeis melanococcus*

melanoxylon *mel-an-oh-ZY-lon*
With black wood, as in *Acacia melanoxylon*

meleagris *mel-EE-uh-gris*
meleagris, meleagre
With spots like a guinea fowl, as in *Fritillaria meleagris*

melliferus *mel-IH-fer-us*
mellifera, melliferum
Producing honey, as in *Euphorbia mellifera*

melliodorus *mel-ee-uh-do-rus*
melliodora, melliodorum
With the scent of honey, as in *Eucalyptus melliodora*

mellitus *mel-IT-tus*
mellita, mellitum
Sweet, like honey, as in *Iris mellita*

meloformis *mel-OH-for-mis*
meloformis, meloforme
Shaped like a melon, as in *Euphorbia meloforme*

membranaceus *mem-bran-AY-see-us*
membranacea, membranaceum
Like skin or membrane, as in *Scadoxus membranaceus*

meniscifolius *men-is-ih-FOH-lee-us*
meniscifolia, meniscifolium
With leaves that are crescent-shaped, as in *Serpocaulon meniscifolium*

menziesii *menz-ESS-ee-eye*
Named after Archibald Menzies (1754–1842), British naval
surgeon and botanist, as in *Pseudotsuga menziesii*

meridianus *mer-id-ee-AH-nus*
meridiana, meridianum
meridionalis *mer-id-ee-oh-NAH-lis*
meridionalis, meridionale
Flowering at noon, as in *Primula* × *meridiana*

metallicus *meh-TAL-ih-kus*
metallica, metallicum
Metallic, as in *Begonia metallica*

mexicanus *meks-sih-KAH-nus*
mexicana, mexicanum
Connected with Mexico, as in *Agastache mexicana*

michauxioides *miss-SHOW-ee-uh-deez*
Resembling *Michauxia*, as in *Campanula michauxioides*

micro-
Used in compound words to denote small

micracanthus *mik-ra-KAN-thus*
micracantha, micracanthum
With small thorns, as in *Euphorbia micracantha*

micranthus *mi-KRAN-thus*
micrantha, micranthum
With small flowers, as in *Heuchera micrantha*

microcarpus *my-kro-KAR-pus*
microcarpa, microcarpum
With small fruit, as in × *Citrofortunella microcarpa*

microcephalus *my-kro-SEF-uh-lus*
microcephala, microcephalum
With a small head, as in *Persicaria microcephala*

microdasys *my-kro-DAS-is*
Small and shaggy, as in *Opuntia microdasys*

microdon *my-kro-DON*
With small teeth, as in *Asplenium* × *microdon*

microglossus *mak-roh-GLOS-us*
microglossa, microglossum
Small-tongued, as in *Ruscus* × *microglossum*

micropetalus *my-kro-PET-uh-lus*
micropetala, micropetalum
With small petals, as in *Cuphea micropetala*

microphyllus *my-kro-FIL-us*
microphylla, microphyllum
With small leaves, as in *Sophora microphylla*

micropterus *mik-rop-TER-us*
microptera, micropterum
With small wings, as in *Promenaea microptera*

Promenaea microptera

microsepalus *mik-ro-SEP-a-lus*
microsepala, microsepalum
With small sepals, as in *Pentadenia microsepala*

miliaceus *mil-ee-AY-see-us*
miliacea, miliaceum
Relating to millet, as in *Panicum miliaceum*

militaris *mil-ih-TAH-ris*
militaris, militare
Relating to soldiers; like a soldier, as in *Orchis militaris*

millefoliatus *mil-le-foh-lee-AH-tus*
millefoliata, millefoliatum
millefolius *mil-le-FOH-lee-us*
millefolia, millefolium
With many leaves (literally a thousand leaves), as in
Achillea millefolium

Castilleja miniata,
giant red paintbrush

mimosoides *mim-yoo-SOY-deez*
Resembling *Mimosa*, as in *Caesalpinia mimosoides*

miniatus *min-ee-AH-tus*
miniata, miniatum
Cinnabar-red, as in *Clivia miniata*

minimus *MIN-eh-mus*
minima, minimum
Smallest, as in *Myosurus minimus*

minor *MY-nor*
minor, minus
Smaller, as in *Vinca minor*

minutus *min-YOO-tus*
minuta, minutum
Very small, as in *Tagetes minuta*

minutiflorus *min-yoo-tih-FLOR-us*
minutiflora, minutiflorum
With minute flowers, as in *Narcissus minutiflorus*

minutifolius *min-yoo-tih-FOH-lee-us*
minutifolia, minutifolium
With minute leaves, as in *Rosa minutifolia*

minutissimus *min-yoo-TEE-sih-mus*
minutissima, minutissimum
The most minute, as in *Primula minutissima*

mirabilis *mir-AH-bih-lis*
mirabilis, mirabile
Wonderful; remarkable, as in *Puya mirabilis*

missouriensis *miss-oor-ee-EN-sis*
missouriensis, missouriense
From Missouri, USA, as in *Iris missouriensis*

mitis *MIT-is*
mitis, mite
Mild; gentle; without spines, as in *Caryota mitis*

mitratus *my-TRAH-tus*
mitrata, mitratum
With a turban or mitre, as in *Mitrophyllum mitratum*

mitriformis *mit-ri-FOR-mis*
mitriformis, mitriforme
Like a cap, as in *Aloe mitriformis*

mixtus *MIKS-tus*
mixta, mixtum
Mixed, as in *Potentilla* × *mixta*

modestus *mo-DES-tus*
modesta, modestum
Modest, as in *Aglaonema modestum*

moesiacus *mee-shee-AH-kus*
moesiaca, moesiacum
Connected with Moesia, the Balkans, as in
Campanula moesiaca

moldavicus *mol-DAV-ih-kus*
moldavica, moldavicum
From Moldavia, eastern Europe, as in *Dracocephalum moldavica*

mollis *MAW-lis*
mollis, molle
Soft; with soft hairs, as in *Alchemilla mollis*

mollissimus *maw-LISS-ih-mus*
mollissima, mollissimum
Very soft, as in *Passiflora mollissima*

moluccanus *mol-oo-KAH-nus*
moluccana, moluccanum
Connected with the Moluccas or Spice Islands, Indonesia,
as in *Pittosporum moluccanum*

monacanthus *mon-ah-KAN-thus*
monacantha, monacanthum
With one spine, as in *Rhipsalis monacantha*

monadelphus *mon-ah-DEL-fus*
monadelpha, monadelphum
With filaments united, as in *Dianthus monadelphus*

monandrus *mon-AN-drus*
monandra, monandrum
With one stamen, as in *Bauhinia monandra*

monensis *mon-EN-sis*
monensis, monense
From Mona, either the Isle of Man or Anglesey, as in
Coincya monensis

mongolicus *mon-GOL-ih-kus*
mongolica, mongolicum
Connected with Mongolia, as in *Quercus mongolica*

Hamamelis mollis is one of the loveliest witch hazels,
with sweetly smelling golden-yellow flowers that make
a welcome appearance in midwinter. *Mollis*, meaning
soft or with soft hairs, refers to the leaves, which are
felted and turn to a glowing yellow in autumn.

Hamamelis mollis,
witch hazel

moniliferus *mon-ih-IH-fer-us*
monilifera, moniliferum
With a necklace, as in *Chrysanthemoides monilifera*

moniliformis *mon-il-lee-FOR-mis*
moniliformis, moniliforme
Like a necklace; with structures resembling strings of beads,
as in *Melpomene moniliformis*

mono-
Used in compound words to denote single

LATIN IN ACTION

Arnica is a genus of perennial and herbaceous plants native to mountainous parts of Europe and has long been used in herbal medicine for pain relief. Its name is sometimes derived from the Greek *arnakis*, sheepskin, in reference to the plant's soft, hairy leaves.

Arnica montana,
leopard's bane

monogynus *mon-NO-gy-nus*
monogyna, monogynum
With one pistil, as in *Crataegus monogyna*

monopetalus *mon-no-PET-uh-lus*
monopetala, monopetalum
With a single petal, as in *Limoniastrum monopetalum*

monophyllus *mon-oh-FIL-us*
monophylla, monophyllum
With one leaf, as in *Pinus monophylla*

monopyrenus *mon-NO-py-ree-nus*
monopyrena, monopyrenum
With a single stone, as in *Cotoneaster monopyrenus*

monostachyus *mon-oh-STAK-ee-us*
monostachya, monostachyum
With one spike, as in *Guzmania monostachya*

monspessulanus *monz-pess-yoo-LAH-nus*
monspessulana, monspessulanum
Connected with Montpellier, France, as in *Acer monspessulanum*

monstrosus *mon-STROH-sus*
monstrosa, monstrosum
Abnormal, as in *Gypsophila × monstrosa*

montanus *MON-tah-nus*
montana, montanum
Relating to mountains, as in *Clematis montana*

montensis *mont-EN-sis*
montensis, montense
monticola *mon-TIH-koh-luh*
Growing on mountains, as in *Halesia monticola*

montigenus *mon-TEE-gen-us*
montigena, montigenum
Born of the mountains, as in *Picea montigena*

morifolius *mor-ee-FOH-lee-us*
morifolia, morifolium
With leaves like the mulberry (*Morus*), as in *Passiflora morifolia*

moschatus *MOSS-kuh-tus*
moschata, moschatum
Musky, as in *Malva moschata*

mucosus *moo-KOZ-us*
mucosa, mucosum
Slimy, as in *Rollinia mucosa*

mucronatus *muh-kron-AH-tus*
mucronata, mucronatum
With a point, as in *Gaultheria mucronata*

mucronulatus *mu-kron-yoo-LAH-tus*
mucronulata, mucronulatum
With a sharp, hard point, as in *Rhododendron mucronulatum*

multi-
Used in compound words to denote many

multibracteatus *mul-tee-brak-tee-AH-tus*
multibracteata, multibracteatum
With many bracts, as in *Rosa multibracteata*

multicaulis *mul-tee-KAW-lis*
multicaulis, multicaule
With many stems, as in *Salvia multicaulis*

multiceps *MUL-tee-seps*
With many heads, as in *Gaillardia multiceps*

multicolor *mul-tee-kol-or*
Multicoloured, as in *Echeveria multicolor*

multicostatus *mul-tee-koh-STAH-tus*
multicostata, multicostatum
With many ribs, as in *Echinofossulocactus multicostatus*

multifidus *mul-TIF-id-us*
multifida, multifidum
With many divisions, usually of leaves with many tears,
as in *Helleborus multifidus*

multiflorus *mul-tih-FLOR-us*
multiflora, multiforum
With many flowers, as in *Cytisus multiflorus*

multilineatus *mul-tee-lin-ee-AH-tus*
multilineata, multilineatum
With many lines, as in *Hakea multilineata*

multinervis *mul-tee-NER-vis*
multinervis, multinerve
With many nerves, as in *Quercus multinervis*

multiplex *MUL-tih-pleks*
With many folds, as in *Bambusa multiplex*

multiradiatus *mul-ty-rad-ee-AH-tus*
multiradiata, multiradiatum
With many rays, as in *Pelargonium multiradiatum*

multisectus *mul-tee-SEK-tus*
multisecta, multisectum
With many cuts, as in *Geranium multisectum*

mundulus *mun-DYOO-lus*
mundula, mundulum
Trim; neat, as in *Gaultheria mundula*

muralis *mur-AH-lis*
muralis, murale
Growing on walls, as in *Cymbalaria muralis*

muricatus *mur-ee-KAH-tus*
muricata, muricatum
With rough and hard points, as in *Solanum muricatum*

musaicus *moh-ZAY-ih-kus*
musaica, musaicum
Like a mosaic, as in *Guzmania musaica*

muscipula *musk-IP-yoo-luh*
Catches flies, as in *Dionaea muscipula*

Cymbalaria muralis,
Solomon's seal

muscivorus *mus-SEE-ver-us*
muscivora, muscivorum
Appearing to eat flies, as in *Helicodiceros muscivorus*

muscoides *mus-COY-deez*
Resembling moss, as in *Saxifraga muscoides*

muscosus *muss-KOH-sus*
muscosa, muscosum
Like moss, as in *Selaginella muscosa*

mutabilis *mew-TAH-bih-lis*
mutabilis, mutabile
Changeable, particularly relating to colour, as in
Hibiscus mutabilis

mutatus *moo-TAH-tus*
mutata, mutatum
Changed, as in *Saxifraga mutata*

muticus *MU-tih-kus*
mutica, muticum
Blunt, as in *Pycnanthemum muticum*

mutilatus *mew-til-AH-tus*
mutilata, mutilatum
Divided as though by tearing, as in *Peperomia mutilata*

myri-
Used in compound words to denote very many

myriacanthus *mir-ee-uh-KAN-thus*
myriacantha, myriacanthum
With very many thorns, as in *Aloe myriacantha*

myriocarpus *mir-ee-oh-KAR-pus*
myriocarpa, myriocarpum
With very many fruits, as in *Schefflera myriocarpa*

myriophyllus *mir-ee-oh-FIL-us*
myriophylla, myriophyllum
With very many leaves, as in *Acaena myriophylla*

myriostigma *mir-ee-oh-STIG-muh*
With many spots, as in *Astrophytum myriostigma*

myrmecophilus *mir-me-koh-FIL-us*
myrmecophila, myrmecophilum
Ant-loving, as in *Aeschynanthus myrmecophilus*

Hibiscus mutabilis,
Confederate rose

myrsinifolius *mir-sin-ee-FOH-lee-us*
myrsinifolia, myrsinifolium
With leaves like *Myrsine*, often referring to myrtle for which this is an
ancient Greek name, as in *Salix myrsinifolia*

myrsinites *mir-SIN-ih-teez*
myrsinoides *mir-sy-NOY-deez*
Resembling *Myrsine*, as in *Gaultheria myrsinoides*

myrtifolius *mir-tih-FOH-lee-us*
myrtifolia, myrtifolium
With leaves like *Myrsine*, as in *Leptospermum myrtifolium*

N

nanellus *nan-EL-lus*
nanella, nanellum
Very dwarf, as in *Lathyrus odoratus* var. *nanellus*

nankingensis *nan-king-EN-sis*
nankingensis, nankingense
From Nanking (Nanjing), China, as in *Chrysanthemum nankingense*

nanus *NAH-nus*
nana, nanum
Dwarf, as in *Betula nana*

napaulensis *nap-awl-EN-sis*
napaulensis, napaulense
From Nepal, as in *Meconopsis napaulensis*

napellus *nap-ELL-us*
napella, napellum
Like a little turnip, referring to the roots, as in *Aconitum napellus*

napifolius *nap-ih-FOH-lee-us*
napifolia, napifolium
With leaves shaped like a turnip (*Brassica rapa*), i.e. a flattened sphere, as in *Salvia napifolia*

narbonensis *nar-bone-EN-sis*
narbonensis, narbonense
From Narbonne, France, as in *Linum narbonense*

narcissiflorus *nar-sis-si-FLOR-us*
narcissiflora, narcissiflorum
With flowers like daffodil (*Narcissus*), as in *Iris narcissiflora*

natalensis *nuh-tal-EN-sis*
natalensis, natalense
From Natal, South Africa, as in *Tulbaghia natalensis*

natans *NAT-anz*
Floating, as in *Trapa natans*

nauseosus *naw-see-OH-sus*
nauseosa, nauseosum
Causing nausea, as in *Chrysothamnus nauseosus*

navicularis *nav-ik-yoo-LAH-ris*
navicularis, naviculare
Shaped like a boat, as in *Callisia navicularis*

LATIN IN ACTION

Originating from Europe and Asia, *Aconitum napellus* is a hardy herbaceous perennial that will reach over 1 m (3 ft) in height if given favourable conditions. Plant them in rich moist soil, preferably in partial shade, although they will tolerate full sun if kept well watered. With their stately and elegant stems bearing dark blue or dark purple hooded flowers, they are great favourites with flower arrangers. However, care should be taken when handling monkshoods, as all parts are poisonous; it is said that hunters used to apply compounds made from the plant to the tips of their arrows to ensure a kill. Indeed, the Romans banned the growing of *A. napellus* and the penalty for disobeying the rule was death. Despite these properties, *A. napellus* is still used in Chinese herbal medicine and in homeopathy.

Aconitum napellus,
monkshood

neapolitanus *nee-uh-pol-ih-TAH-nus*
neapolitana, neapolitanum
Connected with Naples, Italy, as in *Allium neapolitanum*

nebulosus *neb-yoo-LOH-sus*
nebulosa, nebulosum
Like a cloud, as in *Aglaonema nebulosum*

neglectus *nay-GLEK-tus*
neglecta, neglectum
Previously neglected, as in *Muscari neglectum*

nelumbifolius *nel-um-bee-FOH-lee-us*
nelumbifolia, nelumbifolium
With leaves like lotus (*Nelumbo*), as in *Ligularia nelumbifolia*

nemoralis *nem-or-RAH-lis*
nemoralis, nemorale
nemorosus *nem-or-OH-sus*
nemorosa, nemorosum
Of woodland, as in *Anemone nemorosa*

nepalensis *nep-al-EN-sis*
nepalensis, nepalense
nepaulensis *nep-al-EN-sis*
nepaulensis, nepaulense
From Nepal, as in *Hedera nepalensis*

LATIN IN ACTION

Nivalis means white as snow, and this flower's appearance is seen as a sign of hope, as it heralds the passing of sorrow. One theory claims that an angel turned falling snowflakes into the flowers to comfort Adam and Eve after their expulsion from Eden.

Galanthus nivalis, snowdrop

---

nepetoides *nep-et-OY-deez*
Resembling catmint (*Nepeta*), as in *Agastache nepetoides*

neriifolius *ner-ih-FOH-lee-us*
neriifolia, neriifolium
With leaves like oleander (*Nerium*), as in *Podocarpus neriifolius*

nervis *NERV-is*
nervis, nerve
nervosus *ner-VOH-sus*
nervosa, nervosum
With visible nerves, as in *Astelia nervosa*

nicaeensis *ny-see-EN-sis*
nicaeensis, nicaeense
From Nice, France, as in *Acis nicaeensis*

nictitans *Nic-tih-tanz*
Blinking; moving, as in *Chamaecrista nictitans*

nidus *NID-us*
Like a nest, as in *Asplenium nidus*

niger *NY-ger*
nigra, nigrum
Black, as in *Phyllostachys nigra*

nigricans *ny-grih-kanz*
Blackish, as in *Salix nigricans*

nigratus *ny-GRAH-tus*
nigrata, nigratum
Blackened, blackish, as in

nigrescens *ny-GRESS-enz*
Turning black, as in *Silene nigrescens*

nikoensis *nik-o-en-sis*
nikoensis, nikoense
From Nike, Japan, as in *Adenophora nikoensis*

niloticus *nil-OH-tih-kus*
nilotica, niloticum
Connected with the Nile Valley, as in *Salvia nilotica*

nipponicus *nip-PON-ih-kus*
nipponica, nipponicum
Connected with Japan (Nippon), as in *Phyllodoce nipponica*

nitens *NI-tenz*
nitidus *NI-ti-dus*
nitida, nitidum
Shining, as in *Lonicera nitida*

nivalis *niv-VAH-lis*
nivalis, nivale
niveus *NIV-ee-us*
nivea, niveum
nivosus *niv-OH-sus*
nivosa, nivosum
As white as snow, or growing near snow, as in *Galanthus nivalis*

nobilis *NO-bil-is*
nobilis, nobile
Noble; renowned, as in *Laurus nobilis*

noctiflorus *nok-tee-FLOR-us*
noctiflora, noctiflorum
nocturnus *NOK-ter-nus*
nocturna, nocturnum
Flowering at night, as in *Silene noctiflora*

nodiflorus *no-dee-FLOR-us*
nodiflora, nodiflorum
Flowering at the nodes, as in *Eleutherococcus nodiflorus*

nodosus *nod-OH-sus*
nodosa, nodosum
With conspicuous joints or nodes, as in *Geranium nodosum*

nodulosus *no-du-LOH-sus*
nodulosa, nodulosum
With small nodes, as in *Echeveria nodulosa*

noli-tangere *NO-lee TAN-ger-ee*
'Touch not' (because the seed pods burst), as in *Impatiens noli-tangere*

non-scriptus *non-SKRIP-tus*
non-scripta, non-scriptum
Without any markings, as in *Hyacinthoides non-scripta*

norvegicus *nor-VEG-ih-kus*
norvegica, norvegicum
Connected with Norway, as in *Arenaria norvegica*

notatus *no-TAH-tus*
notata, notatum
With spots or marks, as in *Glyceria notata*

novae-angliae *NO-vee ANG-lee-a*
Connected with New England, USA, as in *Aster novae-angliae*

novae-zelandiae *NO-vay zee-LAN-dee-ay*
Connected with New Zealand, as in *Acaena novae-zelandiae*

novi-
Used in compound words to denote new

novi-belgii *NO-vee BEL-jee-eye*
Connected with New York, USA, as in *Aster novi-belgii*

nubicola *noo-BIH-koh-luh*
Growing up in the clouds, as in *Salvia nubicola*

nubigenus *noo-bee-GEE-nus*
nubigena, nubigenum
Born up in the clouds, as in *Kniphofia nubigena*

nucifer *NOO-siff-er*
nucifera, nuciferum
Producing nuts, as in *Cocos nucifera*

nudatus *noo-DAH-tus*
nudata, nudatum
nudus *NEW-dus*
nuda, nudum
Bare; naked, as in *Nepeta nuda*

nudicaulis *new-dee-KAW-lis*
nudicaulis, nudicaule
With bare stems, as in *Papaver nudicaule*

nudiflorus *noo-dee-FLOR-us*
nudiflora, nudiflorum
With flowers that appear before the leaves, as in *Jasminum nudiflorum*

numidicus *nu-MID-ih-kus*
numidica, numidicum
Connected with Algeria, as in *Abies numidica*

nummularius *num-ew-LAH-ree-us*
nummularia, nummularium
Like coins, as in *Lysimachia nummularia*

nutans *NUT-anz*
Nodding, as in *Billbergia nutans*

nyctagineus *nyk-ta-JEE-nee-us*
nyctaginea, nyctagineum
Flowering at night, as in *Mirabilis nyctaginea*

nymphoides *nym-FOY-deez*
Resembling water lily (*Nymphaea*), as in *Hydrocleys nymphoides*

Plants: Their Fragrance and Taste

Smell is probably the most subjective of all the senses, and this may be why some of the Latin terms relating to the fragrance of plants and flowers might appear to be rather misleading. Numerous species names allude to smell, but they should be treated with a degree of caution. For instance, one should not be put off growing the lovely winter-flowering perennial *Helleborus foetidus*, commonly known as stinking hellebore as, despite its name, it only releases an unpleasant smell when crushed. (*Foetidus, foetida, foetidum*, meaning a bad smell.) If you want a scented orchid to grow indoors, do not overlook *Dendrobium anosmum* as, contrary to its appellation, it does in fact have a very strong smell

Rosa foetida,
Austrian briar

Foetida can indicate a plant has a bad smell and many think the fragrance of this rose resembles boiled linseed oil.

(*anosmus, anosma, anosmum*, without scent). With this proviso in mind, if you are searching for plants with a good smell then it would be prudent to start with terms such as *aromaticus* (*aromatica, aromaticum*), meaning fragrant or aromatic; *fragrans* and *fragrantissimus* (*fragrantissima, fragrantissimum*), meaning fragrant and very fragrant; or *odoratissimus* (*odoratissima, odoratissimum*), again meaning very fragrant; or *suaveolens*, which indicates that a plant has a sweet scent. *Graveolens* denotes a plant that has a very pronounced smell, such as *Pelargonium graveolens*, which is often referred to as the sweet-scented geranium. The various species of this plant have a variety of fragrances, many of which are released once the leaves are lightly bruised between the fingers.

Inodorus (*inodora, inodorum*) means a plant without a smell, as in *Hypericum inodorum* while *olidus* (*olida, olidum*) alerts one to an unpleasant fragrance, as in *Eucalyptus olida*. Species names may refer to fragrant parts of the plant other than the flower, such as the fruit or leaf. The evergreen tree *Cinnamomum aromaticum* is prized for its aromatic bark, while the leaves, flowers and seeds of *Myrrhis odorata*, sweet cicely, are all used in cooking, especially in things like tarts and cakes. (*Odoratus, odorata, odoratum*, meaning fragrant.) One is unlikely to find a plant label bearing the legend *zibethinus*, meaning smelling as foul as a civet cat, other than attached to the famous Indonesian fruit durian. It boasts the name *Durio zibethinus*; its fruit tastes sweet and spicy, yet the very pungent smell repels many people, even though the fruit is edible. Some scent terms are quite specific, such as *anisatus* (*anisata, anisatum*), referring to the scent of anise. The Japanese star anise, *Illicium anisatum*, is so pungent that it is burnt as incense. *Caryophyllus*

(*caryophyllus, caryophyllum*) relates to the scent of cloves, as in the much-loved *Dianthus caryophylla*, the clove pink. Many terms that apply to taste helpfully act as something of a warning to the unwary. One term to note and avoid is *emeticus* (*emetica, emeticum*), meaning to cause vomiting. *Acerbus* (*acerba, acerbum*), *amarellus* (*amarella, amarellum*) and *amarus* (*amara, amarum*) all mean

Illicium anisatum,
Japanese star anise

Although highly toxic if consumed, when burnt as incense it is prized for its aromatic qualities.

Lathyrus odoratus,
sweet pea

As the species name suggests, many varieties of sweet pea are among the best scented flowers for the garden.

bitter or sour-tasting, as in the name for the bitter sneezeweed *Helenium amarum*. Another term to be aware of is *causticus* (*caustica, causticum*); this warns that a plant has a burning and caustic effect in the mouth. There are several terms for acidic or sour tastes, including *acidosus* (*acidosa, acidosum*) and *acidus* (*acida, acidum*). By welcome comparison, *dulcis* (*dulcis, dulce*) means sweet, while *sapidus* (*sapida, sapidum*) means a pleasant taste. For something as sweet as honey, try *mellitus*, while *melliodorus* means honey-scented (*mellita, mellitum* and *melliodora, melliodorum*).

O

obconicus *ob-KON-ih-kus*
obconica, obconicum
In the shape of an inverted cone, as in *Primula obconica*

obesus *oh-BEE-sus*
obesa, obesum
Fat, as in *Euphorbia obesa*

oblatus *ob-LAH-tus*
oblata, oblatum
With flattened ends, as in *Syringa oblata*

obliquus *oh-BLIK-wus*
obliqua, obliquum
Lopsided, as in *Nothofagus obliqua*

Cyrtanthus obliquus,
sore-eye flower

oblongatus *ob-long-GAH-tus*
oblongata, oblongatum
oblongus *ob-LONG-us*
oblonga, oblongum
Oblong, as in *Passiflora oblongata*

oblongifolius *ob-long-ih-FOH-lee-us*
oblongifolia, oblongifolium
With oblong leaves, as in *Asplenium oblongifolium*

obovatus *ob-oh-VAH-tus*
obovata, obovatum
In the shape of an inverted egg, as in *Paeonia obovata*

obscurus *ob-SKEW-rus*
obscura, obscurum
Not clear or certain, as in *Digitalis obscura*

obtectus *ob-TEK-tus*
obtecta, obtectum
Covered; protected, as in *Cordyline obtecta*

obtusatus *ob-tew-SAH-tus*
obtusata, obtusatum
Blunt, as in *Asplenium obtusatum*

obtusifolius *ob-too-sih-FOH-lee-us*
obtusifolia, obtusifolium
With blunt leaves, as in *Peperomia obtusifolia*

obtusus *ob-TOO-sus*
obtusa, obtusum
Blunt, as in *Chamaecyparis obtusa*

obvallatus *ob-val-LAH-tus*
obvallata, obvallatum
Enclosed, within a wall, as in *Saussurea obvallata*

occidentalis *ok-sih-den-TAH-lis*
occidentalis, occidentale
Relating to the West, as in *Thuja occidentalis*

occultus *ock-ULL-tus*
occulta, occultum
Hidden, as in *Huernia occulta*

ocellatus *ock-ell-AH-tus*
ocellata, ocellatum
With an eye; with a spot surrounding a smaller spot of a
different colour, as in *Convolvulus ocellatus*

ochraceus *oh-KRA-see-us*
ochracea, ochraceum
Ochre-coloured, as in *Hebe ochracea*

ochroleucus *ock-roh-LEW-kus*
ochroleuca, ochroleucum
Yellowish white, as in *Crocus ochroleucus*

oct-
Used in compound words to denote eight

octandrus *ock-TAN-drus*
octandra, octandum
With eight stamens, as in *Phytolacca octandra*

octopetalus *ock-toh-PET-uh-lus*
octopetala, octopetalum
With eight petals, as in *Dryas octopetala*

oculatus *ock-yoo-LAH-tus*
oculata, oculatum
With an eye, as in *Haworthia oculata*

oculiroseus *ock-yoo-lee-ROH-sus*
oculirosea, oculiroseum
With a rose-coloured eye, as in *Hibiscus palustris* f. *oculiroseus*

ocymoides *ok-kye-MOY-deez*
Resembling basil (*Ocimum*), as in *Halimium ocymoides*

odoratus *oh-dor-AH-tus*
odorata, odoratum
odoriferus *oh-dor-IH-fer-us*
odorifera, odoriferum
odorus *oh-DOR-us*
odora, odorum
With a fragrant scent, as in *Lathyrus odoratus*

odoratissimus *oh-dor-uh-TISS-ih-mus*
odoratissima, odoratissimum
With a very fragrant scent, as in *Viburnum odoratissimum*

officinalis *oh-fiss-ih-NAH-lis*
officinalis, officinale
Sold in shops, hence denoting a useful plant (vegetable, culinary or medicinal herb), as in *Rosmarinus officinalis*

officinarum *off-ik-IN-ar-um*
From a shop, usually an apothecary, as in *Mandragora officinarum*

olbius *OL-bee-us*
olbia, olbium
Connected with the Îles d'Hyères, France, as in *Lavatera olbia*

oleiferus *oh-lee-IH-fer-us*
oleifera, oleiferum
Producing oil, as in *Elaeis oleifera*

oleifolius *oh-lee-ih-FOH-lee-us*
oleifolia, oleifolium
With leaves like olive (*Olea*), as in *Lithodora oleifolia*

LATIN IN ACTION

This wild garlic is found growing in moist, grassy areas across northern Europe and has small and pungent bulbs. *Oleraceum* means from the vegetable garden, but garlic, aside from its culinary uses, has also been used in traditional medicine for centuries.

Allium oleraceum,
field garlic

oleoides *oh-lee-OY-deez*
Resembling olive (*Olea*), as in *Daphne oleoides*

oleraceus *awl-lur-RAY-see-us*
oleracea, oleraceum
Used as a vegetable, as in *Spinacia oleracea*

oliganthus *ol-ig-AN-thus*
oligantha, oliganthum
With few flowers, as in *Ceanothus oliganthus*

oligocarpus *ol-ig-oh-KAR-pus*
oligocarpa, oligocarpum
With few fruits, as in *Cayratia oligocarpa*

oligophyllus *ol-ig-oh-FIL-us*
oligophylla, oligophyllum
With few leaves, as in *Senna oligophylla*

oligospermus *ol-ig-oh-SPERM-us*
oligosperma, oligospermum
With few seeds, as in *Draba oligosperma*

olitorius *ol-ih-TOR-ee-us*
olitoria, olitorium
Relating to culinary herbs, as in
Corchorus olitorius

olivaceus *oh-lee-VAY-see-us*
olivacea, olivaceum
Olive; green-brown, as in *Lithops olivacea*

olympicus *oh-LIM-pih-kus*
olympica, olympicum
Connected with Mount Olympus, Greece, as in
Hypericum olympicum

opacus *oh-PAH-kus*
opaca, opacum
Dark; dull; shaded, as in *Crataegus opaca*

operculatus *oh-per-koo-LAH-tus*
operculata, operculatum
With a cover or lid, as in *Luffa operculata*

ophioglossifolius
oh-fee-oh-gloss-ih-FOH-lee-us
ophioglossifolia, ophioglossifolium
With leaves like adder's tongue fern (*Ophioglossum*),
as in *Ranunculus ophioglossifolius*

oppositifolius *op-po-sih-tih-FOH-lee-us*
oppositifolia, oppositifolium
With leaves that grow opposite each other from the stem, as in
Chiastophyllum oppositifolium

orbicularis *or-bik-yoo-LAH-ris*
orbicularis, orbiculare
orbiculatus *or-bee-kul-AH-tus*
orbiculata, orbiculatum
In the shape of a disc; flat and round, as in *Cotyledon orbiculata*

orchideus *or-KI-de-us*
orchidea, orchideum
orchioides *or-ki-OY-deez*
Like an orchid (*Orchis*), as in *Veronica orchidea*

Chiastophyllum oppositifolium,
lamb's tail

orchidiflorus *or-kee-dee-FLOR-us*
orchidiflora, orchidiflorum
With flowers like an orchid (*Orchis*), as in *Gladiolus orchidiflorus*

oreganus *or-reh-GAH-nus*
oregana, oreganum
Connected with Oregon, USA, as in *Sidalcea oregana*

oreophilus *or-ee-O-fil-us*
oreophila, oreophilum
Mountain-loving, as in *Sarracenia oreophila*

oresbius *or-ES-bee-us*
oresbia, oresbium
Growing on mountains, as in *Castilleja oresbia*

orientalis *or-ee-en-TAH-lis*
orientalis, orientale
Relating to the Orient; Eastern, as in *Thuja orientalis*

origanifolius *or-ih-gan-ih-FOH-lee-us*
origanifolia, origanifolium
With leaves like marjoram (*Origanum*), as in
Chaenorhinum origanifolium

origanoides *or-ig-an-OY-deez*
Resembling marjoram (*Origanum*), as in *Dracocephalum origanoides*

ornans *OR-nanz*
ornatus *or-NA-tus*
ornata, ornatum
Ornamental; showy, as in *Musa ornata*

ornatissimus *or-nuh-TISS-ih-mus*
ornatissima, ornatissimum
Very showy, as in *Bulbophyllum ornatissimum*

ornithopodus *or-nith-OP-oh-dus*
ornithopoda, ornithopodum
ornithopus *or-nith-OP-pus*
Like a bird's foot, as in *Carex ornithopoda*

ortho-
Used in compound words to denote straight or upright

orthobotrys *or-THO-bot-ris*
With upright clusters, as in *Berberis orthobotrys*

orthocarpus *or-tho-KAR-pus*
orthocarpa, orthocarpum
With upright fruit, as in *Malus orthocarpa*

orthoglossus *or-tho-GLOSS-us*
orthoglossa, orthoglossum
With a straight tongue, as in *Bulbophyllum orthoglossum*

orthosepalus *or-tho-SEP-a-lus*
orthosepala, orthosepalum
With straight sepals, as in *Rubus orthosepalus*

osmanthus *os-MAN-thus*
osmantha, osmanthum
With fragrant flowers, as in *Phyllagathis osmantha*

ovalis *oh-VAH-lis*
ovalis, ovale
Oval, as in *Amelanchier ovalis*

ovatus *oh-VAH-tus*
ovata, ovatum
Shaped like an egg; ovate, as in *Lagurus ovatus*

ovinus *oh-VIN-us*
ovina, ovinum
Relating to sheep or sheep feed, as in *Festuca ovina*

oxyacanthus *oks-ee-a-KAN-thus*
oxyacantha, oxyacanthum
With sharp spines, as in *Asparagus oxyacanthus*

oxygonus *ok-SY-goh-nus*
oxygona, oxygonum
With sharp angles, as in *Echinopsis oxygona*

oxyphilus *oks-ee-FIL-us*
oxyphila, oxyphilum
Growing in acid soils, as in *Allium oxyphilum*

oxyphyllus *oks-ee-FIL-us*
oxyphylla, oxyphyllum
With sharp-pointed leaves, as in *Euonymus oxyphyllus*

P

pachy-
Used in compound words to denote thick

pachycarpus *pak-ih-KAR-pus*
pachycarpa, pachycarpum
With a thick pericarp, as in *Angelica pachycarpa*

pachyphyllus *pak-ih-FIL-us*
pachyphylla, pachyphyllum
Thick-leaved, as in *Callistemon pachyphyllus*

pachypodus *pak-ih-POD-us*
pachypoda, pachypodum
With a fat stem, as in *Actaea pachypoda*

pacificus *pa-SIF-ih-kus*
pacifica, pacificum
Connected with the Pacific Ocean, as in *Chrysanthemum pacificum*

pachypterus *pak-IP-ter-us*
pachyptera, pachypterum
With thick wings, as in *Rhipsalis pachyptera*

pachysanthus *pak-ee-SAN-thus*
pachysantha, pachysanthum
With thick flowers, as in *Rhododendron pachysanthum*

padus *PAD-us*
Ancient Greek name for a kind of wild cherry, as in *Prunus padus*

paganus *PAG-ah-nus*
pagana, paganum
From wild or country regions, as in *Rubus paganus*

palaestinus *pal-ess-TEEN-us*
palaestina, palaestinum
Connected with Palestine, as in *Iris palaestina*

pallens *PAL-lenz*
pallidus *PAL-lid-dus*
pallida, pallidum
Pale, as in *Tradescantia pallida*

pallescens *pa-LESS-enz*
Rather pale, as in *Sorbus pallescens*

pallidiflorus *pal-id-uh-FLOR-us*
pallidiflora, pallidiflorum
With pale flowers, as in *Eucomis pallidiflora*

palmaris *pal-MAH-ris*
palmaris, palmare
A hand's breadth wide, as in *Limonium palmare*

palmatus *pahl-MAH-tus*
palmata, palmatum
Palmate, as in *Acer palmatum*

palmensis *pal-MEN-sis*
palmensis, palmense
From Las Palmas, Canary Islands, as in *Aichryson palmense*

palmeri *PALM-er-ee*
Named after Ernest Jesse Palmer (1875–1962), English explorer and plant collector in the United States, as in *Agave palmeri*

LATIN IN ACTION

Originating from Cape Province, South Africa, nerines are half-hardy bulbs, with the exception of *Nerine bowdenii*, a species that can be planted outside in cooler regions. The plant flowers in autumn and some have pale pink blooms, as described by the cultivar name 'Pallida'.

Nerine bowdenii,
Cape or Cornish lily

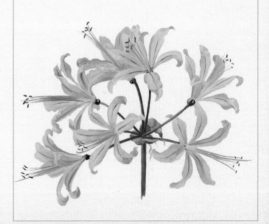

palmetto *pahl-MET-oh*
A small palm, as in *Sabal palmetto*

palmifolius *palm-ih-FOH-lee-us*
palmifolia, palmifolium
With palm-like leaves, as in *Sisyrinchium palmifolium*

paludosus *pal-oo-DOH-sus*
paludosa, paludosum
palustris *pal-US-tris*
palustris, palustre
Of marshland, as in *Quercus palustris*

pandanifolius *pan-dan-uh-FOH-lee-us*
pandanifolia, pandanifolium
With leaves like *Pandanus*, as in *Eryngium pandanifolium*

panduratus *pand-yoor-RAH-tus*
pandurata, panduratum
Shaped like a fiddle, as in *Coelogyne pandurata*

paniculatus *pan-ick-yoo-LAH-tus*
paniculata, paniculatum
With flowers arranged in panicles, as in *Koelreuteria paniculata*

pannonicus *pa-NO-nih-kus*
pannonica, pannonicum
Of Pannonia, a Roman region, as in *Lathyrus pannonicus*

pannosus *pan-OH-sus*
pannosa, pannosum
Tattered, as in *Helianthemum pannosum*

papilio *pap-ILL-ee-oh*
A butterfly, as in *Hippeastrum papilio*

papilionaceus *pap-il-ee-on-uh-SEE-us*
papilionacea, papilionaceum
Like a butterfly, as in *Pelargonium papilionaceum*

papyraceus *pap-ih-REE-see-us*
papyracea, papyraceum
Like paper, as in *Narcissus papyraceus*

papyrifer *pap-IH-riff-er*
papyriferus *pap-ih-RIH-fer-us*
papyrifera, papyriferum
Producing paper, as in *Tetrapanax papyrifer*

papyrus *pa-PY-rus*
Ancient Greek word for paper, as in *Cyperus papyrus*

Anchusa azurea
(syn. *Anchusa paniculata*)

paradisi *par-ih-DEE-see*
paradisiacus *par-ih-DEE-see-cus*
paradisiaca, paradisiacum
From parks or gardens, as in *Citrus × paradisi*

paradoxus *par-uh-DOKS-us*
paradoxa, paradoxum
Unexpected or paradoxical, as in *Acacia paradoxa*

paraguayensis *par-uh-gway-EN-sis*
paraguayensis, paraguayense
From Paraguay, as in *Ilex paraguayensis*

parasiticus *par-uh-SIT-ih-kus*
parasitica, parasiticum
Parasitic, as in *Agalmyla parasitica*

pardalinus *par-da-LEE-nus*
pardalina, pardalinum
pardinus *par-DEE-nus*
pardina, pardinum
With spots like a leopard, as in *Hippeastrum pardinum*

pari-
Used in compound words to denote equal

parnassicus *par-NASS-ih-kus*
parnassica, parnassicum
Connected with Mount Parnassus, Greece, as in *Thymus parnassicus*

parnassifolius *par-nass-ih-FOH-lee-us*
parnassifolia, parnassifolium
With leaves like the grass-of-Parnassus (*Parnassia*), as in
Saxifraga parnassifolia

parryae *PAR-ee-eye*
parryi *PAIR-ree*
Named after Dr Charles Christopher Parry (1823–90), English-born
botanist and plant collector. The form *parryae* commemorates his
wife, Emily Richmond Parry (1821–1915), as in *Linanthus parryae*.

Campanula patula,
spreading bellflower

partitus *par-TY-tus*
partita, partitum
Parted, as in *Hibiscus partitus*

parvi-
Used in compound words to denote small

parviflorus *par-vee-FLOR-us*
parviflora, parviflorum
With small flowers, as in *Aesculus parviflora*

parvifolius *par-vih-FOH-lee-us*
parvifolia, parvifolium
With small leaves, as in *Eucalyptus parvifolia*

parvus *PAR-vus*
parva, parvum
Small, as in *Lilium parvum*

patagonicus *pat-uh-GOH-nih-kus*
patagonica, patagonicum
Connected with Patagonia, as in *Sisyrinchium patagonicum*

patavinus *pat-uh-VIN-us*
patavina, patavinum
Connected with Padua (previously Patavium), Italy, as in
Haplophyllum patavinum

patens *PAT-enz*
patulus *PAT-yoo-lus*
patula, patulum
With a spreading habit, as in *Salvia patens*

pauci-
Used in compound words to denote few

pauciflorus *PAW-ki-flor-us*
pauciflora, pauciflorum
With few flowers, as in *Corylopsis pauciflora*

paucifolius *paw-ke-FOH-lee-us*
paucifolia, paucifolium
With few leaves, as in *Scilla paucifolia*

paucinervis *paw-ke-NER-vis*
paucinervis, paucinerve
With few nerves, as in *Cornus paucinervis*

pauperculus *paw-PER-yoo-lus*
paupercula, pauperculum
Poor, as in *Alstroemeria paupercula*

PARTHENOCISSUS

Virginia creeper, *Parthenocissus quinquefolia*, has a convoluted nomenclatural history. In 1753, Linnaeus named the plant *Hedera quinquefolia*; it was later assigned to *Vitis* (grapevine) and *Ampelopsis*, among other genera, before the French botanist Jules Émile Planchon (1823–1888) reclassified it under its current name in 1887. *Parthenocissus* derives from the Greek *parthenos*, virgin, and *kissos*, meaning ivy, and was devised by Planchon as a direct translation of the French common name for the plant, *vigne-vierge*, first recorded in 1690. This in turn was a reference to the American colony from which *P. quinquefolia* takes its English name. Through translation into French, then Latinised Greek, the botanical name thus indirectly commemorates Queen Elizabeth I, the 'Virgin Queen', in whose honour Virginia was named. More prosaically, *quinquefolius* (*quinquefolia*, *quinquefolium*) means with five leaves and describes the plant's attractive foliage. Similarly, Boston ivy, also known as Japanese creeper, is properly called *P. tricuspidata*, the name referring to its three-lobed leaves (*tricuspidatus*, *tricuspidata*, *tricuspidatum*, with three points). *P. heptaphylla* meanwhile is commonly known as sevenleaf (*heptaphyllus*, *heptaphylla*, *heptaphyllum*, with seven leaves).

Part of the family *Vitaceae*, *Parthenocissus* are quick-growing deciduous climbers that can creep up a wall to a height of 30 m (95 ft). They are famous for their beautiful and long-lasting autumn colour and will gracefully drape over a pergola, fence or tree. All species are self-clinging and suited to walls, producing

This honeysuckle *Lonicera periclymenum* shares its the common name woodbine with *Parthenocissus vitacea*.

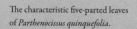

The characteristic five-parted leaves of *Parthenocissus quinquefolia*.

sticky pads at the tips of their tendrils. *Parthenocissus* are grown mainly for their attractive foliage, as their flowers are tiny and insignificant. If grown at the base of a wall or tree, be sure to incorporate plenty of compost when planting and water frequently in the first few seasons until the plant is well established. Be aware that the sap can cause skin irritation in some people.

P. vitacea (*vitaceus*, *vitacea*, *vitaceum* like *Vitis*, vine) is known as thicket creeper, and also woodbine or grape woodbine.

Confusingly, the term woodbine is also often used for other climbers such as *Lonicera periclymenum* (honeysuckle).

PASSIFLORA

Belonging to the family *Passifloraceae*, passion flowers are vigorous evergreen climbers with very striking exotic-looking flowers. There are a great many species and cultivars. It is their large saucer-shaped flowers that inspired the name given to them by early Catholic missionaries working in South America. *Passio* is Latin for passion or suffering, particularly associated with the Passion of Christ, while *flos* means flower. In the eyes of the missionaries, the configuration of the flower head became a visual symbol of the crucifixion; the corona threads represented the crown of thorns; the five sepals and five petals the ten apostles (less Peter, who denied Christ and Judas, who betrayed him); the stigmas recalled the nails of the Passion, and the anthers the wounds they inflicted; the coiled tendrils were likened to the cords used to flog Christ, while the digitate or finger-like leaves resembled the outstretched hands of the multitude. Old European names for the plant continue this theme, and include Christ's thorn and Christ's bouquet.

Passiflora edulis bears edible yellow or purple egg-shaped fruits called passionfruit. *P. edulis* f. *flavicarpa* is grown for its delicious yellow fruit. (*Edulis, edulis, edule* means edible, *flavus* means yellow, and *carpus, carpa, carpum* means fruit.) Native to South America, it is also known as golden

The common blue passionflower, *Passiflora caerulea*, has a complex flower structure.

passionfruit or maracuyá. In cooler regions, hardy passifloras are grown primarily for their flowers. They often bear fruit towards the end of the growing season, but this can cause stomach upsets, especially if not fully ripe. Particularly popular species are *P. caerulea*, the common blue passionflower, and *P. incarnata*. (*Caeruleus, caerulea, caeruleum* means sky-blue; *incarnatus, incarnata, incarnatum*, the colour of flesh.) Both of these are pretty hardy if grown in a sheltered position. Other species may need the protection of a greenhouse or conservatory in cooler climates.

P. alata is the wing-stemmed passionflower; *P. ligularis* is sometimes referred to as sweet granadilla and is grown in Africa and Australia (*alatus, alata, alatum* meaning winged, *ligularis, ligularis, ligulare*, shaped like a strap.)

Passiflora quandrangularis has a curious four-sided stem.

pavia *PAH-vee-uh*
Named after Peter Paaw (1564–1617), Dutch physician, as in
Aesculus pavia

pavoninus *pav-ON-ee-nus*
pavonina, pavoninum
Peacock blue, as in *Anemone pavonina*

pectinatus *pek-tin-AH-tus*
pectinata, pectinatum
Like a comb, as in *Euryops pectinatus*

pectoralis *pek-TOR-ah-lis*
pectoralis, pectorale
Of the chest, as in *Justicia pectoralis*

peculiaris *pe-kew-lee-AH-ris*
peculiaris, peculiare
Peculiar or special, as in *Cheiridopsis peculiaris*

pedatifidus *ped-at-ee-FEE-dus*
pedatifida, pedatifidum
Divided like a bird's foot, as in *Viola pedatifida*

pedatus *ped-AH-tus*
pedata, pedatum
Shaped like a bird's foot, often in reference to the shape of a
palmate leaf, as in *Adiantum pedatum*

pedemontanus *ped-ee-MON-tah-nus*
pedemontana, pedemontanum
Connected with Piedmont, Italy, as in *Saxifraga pedemontana*

peduncularis *pee-dun-kew-LAH-ris*
peduncularis, pedunculare
pedunculatus *pee-dun-kew-LA-tus*
pedunculata, pedunculatum
With a flower stalk, as in *Lavandula pedunculata*

pedunculosus *ped-unk-yoo-LOH-sus*
pedunculosa, pedunculosum
With many or particularly well-developed flower stems, as in
Ilex pedunculosa

pekinensis *pee-keen-EN-sis*
pekinensis, pekinense
From Peking (Beijing), China, as in *Euphorbia pekinensis*

pelegrina *pel-e-GREE-nuh*
Local name for *Alstroemeria pelegrina*

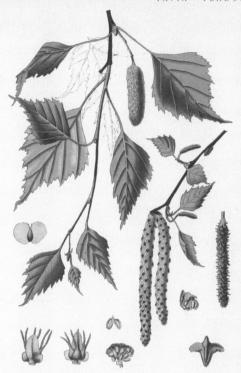

Betula pendula,
silver birch

pellucidus *pel-LOO-sid-us*
pellucida, pellucidum
Transparent; clear, as in *Conophytum pellucidum*

peloponnesiacus *pel-uh-pon-ee-see-AH-kus*
peloponnesiaca, peloponnesiacum
Connected with the Peloponnese, Greece, as in
Colchicum peloponnesiacum

peltatus *pel-TAH-tus*
peltata, peltatum
Shaped like a shield, as in *Darmera peltata*

pelviformis *pel-vih-FORM-is*
pelviformis, pelviforme
In the shape of a shallow cup, as in *Campanula pelviformis*

pendulinus *pend-yoo-LIN-us*
pendulina, pendulinum
Hanging, as in *Salix × pendulina*

pendulus *PEND-yoo-lus*
pendula, pendulum
Hanging, as in *Betula pendula*

penicillatus *pen-iss-sil-LAH-tus*
penicillata, penicillitum
penicillius *pen-iss-SIL-ee-us*
penicillia, penicillium
With a tuft of hair, as in *Parodia penicillata*

peninsularis *pen-in-sul-AH-ris*
peninsularis, peninsulare
From peninsular regions, as in *Allium peninsulare*

penna-marina *PEN-uh mar-EE-nuh*
Sea feather, as in *Blechnum penna-marina*

pennatus *pen-AH-tus*
pennata, pennatum
With feathers, pinnate, as in *Stipa pennata*

pennigerus *pen-NY-ger-us*
pennigera, pennigerum
With leaves like feathers, as in *Thelypteris pennigera*

pennsylvanicus *pen-sil-VAN-ih-kus*
pennsylvanica, pennsylvanicum
pensylvanicus
pensylvanica, pensylvanicum
Connected with Pennsylvania, USA, as in *Acer pensylvanicum*

pensilis *PEN-sil-is*
pensilis, pensile
Hanging, as in *Glyptostrobus pensilis*

penta-
Used in compound words to denote five

pentagonius *pen-ta-GON-ee-us*
pentagonia, pentagonium
pentagonus *pen-ta-GON-us*
pentagona, pentagonum
With five angles, as in *Rubus pentagonus*

pentagynus *pen-ta-GY-nus*
pentagyna, pentagynum
With five pistils, as in *Crataegus pentagyna*

pentandrus *pen-TAN-drus*
pentandra, pentandrum
With five stamens, as in *Ceiba pentandra*

pentapetaloides *pen-ta-pet-al-OY-deez*
Appearing to possess five petals, as in *Convolvulus pentapetaloides*

pentaphyllus *pen-tuh-FIL-us*
pentaphylla, pentaphyllum
With five leaves or leaflets, as in *Cardamine pentaphylla*

pepo *PEP-oh*
Latin word for a large melon or pumpkin, as in *Cucurbita pepo*

perbellus *per-BELL-us*
perbella, perbellum
Very beautiful, as in *Mammillaria perbella*

peregrinus *per-uh-GREE-nus*
peregrina, peregrinum
From the local vernacular name for *Delphinium peregrinum*

perennis *per-EN-is*
perennis, perenne
Perennial, as in *Bellis perennis*

perfoliatus *per-foh-lee-AH-tus*
perfoliata, perfoliatum
With the leaf surrounding the stem, as in *Parahebe perfoliata*

perforatus *per-for-AH-tus*
perforata, perforatum
With, or appearing to have, small holes, as in *Hypericum perforatum*

pergracilis *per-GRASS-il-is*
pergracilis, pergracile
Very slender, as in *Scleria pergracilis*

pernyi *PERN-yee-eye*
Named after Paul Hubert Perny (1818-1907), French missionary
and botanist, as in *Ilex pernyi*

persicifolius *per-sik-ih-FOH-lee-us*
persicifolia, persicifolium
With leaves like the peach (*Prunus persica*), as in
Campanula persicifolia

persicus *PER-sih-kus*
persica, persicum
Connected with Persia (Iran), as in *Parrotia persica*

persistens *per-SIS-tenz*
Persistent, as in *Elegia persistens*

persolutus *per-sol-YEW-tus*
persoluta, persolutum
Very loose, as in *Erica persoluta*

perspicuus *PER-spic-kew-us*
perspicua, perspicuum
Transparent, as in *Erica perspicua*

pertusus *per-TUS-us*
pertusa, pertusum
Perforated; pierced, as in *Listrostachys pertusa*

LATIN IN ACTION

Phaeus **means dusky; in** *Geranium phaeum,* **this relates to the dark and rather sombre flowers and explains its common name. This is a useful hardy geranium, as it is as happy growing in dry shade as full sun. The clump-forming leaves sometimes have attractive purple spots.**

Geranium phaeum,
mourning widow

---

perulatus *per-uh-LAH-tus*
perulata, perulatum
With perules (bud scales), as in *Enkianthus perulatus*

peruvianus *per-u-vee-AH-nus*
peruviana, peruvianum
Connected with Peru, as in *Scilla peruviana*

petaloideus *pet-a-LOY-dee-us*
petaloidea, petaloideum
Like a petal, as in *Thalictrum petaloideum*

petiolaris *pet-ee-OH-lah-ris*
petiolaris, petiolare
petiolatus *pet-ee-oh-LAH-tus*
petiolata, petiolatum
With a leaf stalk, as in *Helichrysum petiolare*

petraeus *pet-RAY-us*
petraea, petraeum
Connected with rocky regions, as in *Quercus petraea*

phaeacanthus *fay-uh-KAN-thus*
phaeacantha, phaeacanthum
With grey thorns, as in *Opuntia phaeacantha*

phaeus *FAY-us*
phaea, phaeum
Dusky, as in *Geranium phaeum*

philadelphicus *fil-uh-DEL-fih-kus*
philadelphica, philadelphicum
Connected with Philadelphia, as in *Lilium philadelphicum*

philippensis *fil-lip-EN-sis*
philippensis, philippense
philippianus *fil-lip-ee-AH-nus*
philippiana, philippianum
philippii *fil-LIP-ee-eye*
philippinensis *fil-ip-ee-NEN-sis*
philippinensis, philippinense
From the Philippines, as in *Adiantum philippense*

phleoides *flee-OY-deez*
Resembling *Phleum* (cat's tail grass), as in *Phleum phleoides*

phlogiflorus *flo-GIF-flor-us*
phlogiflora, phlogiflorum
With flame-coloured flowers or flowers like *Phlox*, as in *Verbena phlogiflora*

JANE COLDEN
(1724–1766)

MARIANNE NORTH
(1830–90)

Sadly, hardly any women appear among the annals of important botanists and plant collectors. It is no coincidence that during the period of the great pioneering plant hunters, the 18th and 19th centuries, very few women indeed received the kind of education that would provide them with that prerequisite knowledge for botanists, a thorough grounding in Greek and Latin. Nor did women travel with any degree of freedom or independence. As a consequence, doubtless many serious botanising women remain unrecorded by history. Fortunately, the American woman Jane Colden and Englishwoman Marianne North are among those who are still remembered for their contributions.

Jane Colden is celebrated as America's first female botanist to use the Linnean system to classify native wildflowers. Her father, Dr Cadwallader Colden, was the Surveyor General of the Colonies and a member of the King's Council of New York. He was a Scotsman with a keen interest in botany. He corresponded with Linnaeus (see p. 132), and his documenting of the plants that grew on his estate west of Newburgh, New York, resulted in the volume *Plantae Coldenhamiae*. Jane Colden received little formal education, but gained some basic knowledge of Latin through her study of plants and of Linnaeus. Acknowledged as a serious collector by contemporary naturalists and botanists, she knew John Bartram (see p. 98), the botanist John Clayton and the London plant collector Peter Collinson. Following correspondence with Dr Alexander Garden of South Carolina (after whom the genus *Gardenia* was named), Colden published a number of scholarly articles on *Hypericum virginicum* (St John's wort).

Married to a Scottish doctor named Farquhar, Jane Colden died aged 42 in 1766, in the same year that her only child also died. Her reputation grew posthumously, and wider recognition came four years after her death when some of her carefully observed plant descriptions were published in the second volume of *Essays and Observations*. Today, one of her manuscripts is held by London's British Museum; it contains more than 300 descriptions of plants, all accompanied by illustrations. The Latin and common names of the plants are also given, along with their flowering time, seeds and medicinal properties.

'SHE DESERVES TO BE CELEBRATED ...
PERHAPS THE FIRST LADY THAT HAS PERFECTLY
STUDIED LINNAEUS'S SYSTEM.'

Peter Collinson, of Jane Colden

Like Jane Colden, Marianne North was educated at home, and she too expressed an early interest in the natural world. As a young woman, she grew various fungi in her room and made frequent trips to London's Kew Gardens to draw and paint rare plant specimens, encouraged by the then director William Hooker (see p. 182). As daughter of the Member of Parliament for Hastings, East Sussex, Frederick North, Marianne was well connected and travelled abroad with her parents, something that she continued to do with her father following the death of her mother. However, her life changed drastically when her adored father died. Aged 40, unmarried and independently wealthy, North set out on a series of exotic and adventurous journeys. She travelled largely alone, a great rarity at that time, saying that her intention was to go 'to some tropical country to paint its peculiar vegetation in its natural abundant luxuriance'.

North's most active period of painting spanned 13 years and included journeys to the United States, Brazil, Canada, Chile, India, Jamaica, Japan, Java, Singapore, South Africa and Tenerife. Her friend Charles Darwin urged her to travel to Australia and New Zealand, a suggestion she readily acted upon. Once abroad, she did not confine herself to painting

North studied briefly under the Victorian flower painter Valentine Bartholomew. This study of *Nepenthes northiana* demonstrates her vigorous style.

native flora and fauna within the safe confines of the world's botanical gardens, but instead journeyed deep into the terrain of each region she visited. North discovered and recorded with great accuracy many plants that were then new to science, sending her oil paintings back to Sir Joseph Hooker (see p. 182) at Kew. The genus *Northia* was named after her, along with species such as *Areca northiana*, *Crinum northianum* and *Kniphofia northiana*. Despite having received little formal artistic training, her work is marked by a vivid use of colour and a fluid handling of paint. The Marianne North Gallery at Kew Gardens opened in 1882 and now houses 832 of her canvases, depicting more than 900 species of plant.

A photograph of North at work painting a Tamil boy, taken by Julia Margaret Cameron at her house in Ceylon (Sri Lanka), 1877.

phoeniceus *feen-ih-KEE-us*
phoenicea, phoeniceum
Purple/red, as in *Juniperus phoenicea*

phoenicolasius *fee-nik-oh-LASS-ee-us*
phoenicolasia, phoenicolasium
With purple hairs, as in *Rubus phoenicolasius*

phrygius *FRIJ-ee-us*
phrygia, phrygium
Connected with Phrygia, Anatolia, as in *Centaurea phrygia*

phyllostachyus *fy-lo-STAY-kee-us*
phyllostachya, phyllostachyum
With a leaf spike, as in *Hypoestes phyllostachya*

picturatus *pik-tur-AH-tus*
picturata, picturatum
With variegated leaves, as in *Calathea picturata*

pictus *PIK-tus*
picta, pictum
Painted; highly coloured, as in *Acer pictum*

pileatus *py-lee-AH-tus*
pileata, pileatum
With a cap, as in *Lonicera pileata*

piliferus *py-LIH-fer-us*
pilifera, piliferum
With short, soft hairs, as in *Ursinia pilifera*

pillansii *pil-AN-see-eye*
Named after Neville Stuart Pillans (1884–1964), South African
botanist, as in *Watsonia pillansii*

pilosus *pil-OH-sus*
pilosa, pilosum
With long, soft hairs, as in *Aster pilosus*

pilularis *pil-yoo-LAH-ris*
pilularis, pilulare
piluliferus *pil-loo-LIH-fer-us*
pilulifera, piluliferum
With globular fruit, as in *Urtica pilulifera*

pimeleoides *py-mee-lee-OY-deez*
Resembling *Pimelea*, as in *Pittosporum pimeleoides*

pimpinellifolius *pim-pi-nel-ih-FOH-lee-us*
pimpinellifolia, pimpinellifolium
With leaves like anise (*Pimpinella*), as in *Rosa pimpinellifolia*

pinetorum *py-net-OR-um*
Connected with pine forests, as in *Fritillaria pinetorum*

pineus *PY-nee-us*
pinea, pineum
Relating to pine (*Pinus*), as in *Pinus pinea*

pinguifolius *pin-gwih-FOH-lee-us*
pinguifolia, pinguifolium
With fat leaves, as in *Hebe pinguifolia*

pinifolius *pin-ih-FOH-lee-us*
pinifolia, pinifolium
With leaves like pine (*Pinus*), as in *Penstemon pinifolius*

pininana *pin-in-AH-nuh*
A dwarf pine, as in *Echium pininana*

pinnatus *pin-NAH-tus*
pinnata, pinnatum
With leaves that grow from each side of a stalk; like a feather,
as in *Santolina pinnata*

pinnatifidus *pin-nat-ih-FY-dus*
pinnatifida, pinnatifidum
Cut in the form of a feather, as in *Eranthis pinnatifida*

pinnatifolius *pin-nat-ih-FOH-lee-us*
pinnatifolia, pinnatifolium
With leaves like feathers, as in *Meconopsis pinnatifolia*

pinnatifrons *pin-NAT-ih-fronz*
With fronds like feathers, as in *Chamaedorea pinnatifrons*

piperitus *pip-er-EE-tus*
piperita, piperitum
With a pepper-like taste, as in *Mentha × piperita*

pisiferus *pih-SIH-fer-us*
pisifera, pisiferum
Bearing peas, as in *Chamaecyparis pisifera*

pitardii *pit-ARD-ee-eye*
Named after Charles-Joseph Marie Pitard-Briau, 20th-century
French plant collector and botanist, as in *Camellia pitardii*

pittonii *pit-TON-ee-eye*
Named after Josef Claudius Pittoni, 19th-century Austrian botanist,
as in *Sempervivum pittonii*

LATIN IN ACTION

Acer platanoides is a fast-growing and vigorous tree that reaches a considerable size. (*A. platanoides* 'Drummondii' is less vigorous and more suited to small gardens.) A robust tree, it is useful for planting as a windbreak, and is hardy and deciduous. Many acers have insignificant flowers, but *A. platanoides* has attractive bright yellow-green flowers and provides good autumn foliage. Its botanical name, *platanoides,* tells us that it resembles the plane tree *Platanus*, while the famous London plane tree is called *Platanus* × *acerifolia*, meaning it resembles the maple (*Acer*). Further confusion arises from the London plane's common name, as it is not in fact a native of Britain. It has become identified with the capital city as it thrives so well in its streets, having a much greater tolerance to traffic pollution than many other trees.

Acer platanoides,
Norway maple

planus *PLAH-nus*
plana, planum
Flat, as in *Eryngium planum*

planiflorus *plen-ee-FLOR-us*
planiflora, planiflorum
With flat flowers, as in *Echidnopsis planiflora*

planifolius *plan-ih-FOH-lee-us*
planifolia, planifolium
With flat leaves, as in *Iris planifolia*

planipes *PLAN-ee-pays*
With a flat stalk, as in *Euonymus planipes*

plantagineus *plan-tuh-JIN-ee-us*
plantaginea, plantagineum
Like plantain (*Platago*), as in *Hosta plantaginea*

platanifolius *pla-tan-ih-FOH-lee-us*
platanifolia, platanifolium
With leaves like a plane tree (*Platanus*), as in *Begonia platanifolia*

platanoides *pla-tan-OY-deez*
Resembling a plane tree (*Platanus*), as in *Acer platanoides*

platy-
Used in compound words to denote broad (or sometimes flat)

platycanthus *plat-ee-KAN-thus*
platycantha, platycanthum
With broad spines, as in *Acaena platycantha*

platycarpus *plat-ee-KAR-pus*
platycarpa, platycarpum
With broad fruit, as in *Thalictrum platycarpum*

platycaulis *plat-ee-KAWL-is*
platycaulis, platycaule
With a broad stem, as in *Allium platycaule*

platycladus *plat-ee-KLAD-us*
platyclada, platycladum
With flat branches, as in *Euphorbia platyclada*

platyglossus *plat-ee-GLOSS-us*
platyglossa, platyglossum
With a broad tongue, as in *Phyllostachys platyglossa*

platypetalus *plat-ee-PET-uh-lus*
platypetala, platypetalum
With broad petals, as in *Epimedium platypetalum*

platyphyllos *plat-tih-FIL-los*
platyphyllus *pla-tih-FIL-us*
platyphylla, platyphyllum
With broad leaves, as in *Betula platyphylla*

platypodus *pah-tee-POD-us*
platypoda, platypodum
With a broad stalk, as in *Fraxinus platypoda*

platyspathus *plat-ees-PATH-us*
platyspatha, platyspathum
With a broad spathe, as in *Allium platyspathum*

platyspermus *plat-ee-SPER-mus*
platysperma, platyspermum
With broad seeds, as in *Hakea platysperma*

pleniflorus *plen-ee-FLOR-us*
pleniflora, pleniflorum
With double flowers, as in *Kerria japonica* 'Pleniflora'

Aegopodium podagraria,
ground elder

plenissimus *plen-ISS-i-mus*
plenissima, plenissimum
Very double-flowered, as in *Eucalyptus kochii* subsp. *plenissima*

plenus *plen-US*
plena, plenum
Double; full, as in *Felicia plena*

plicatus *ply-KAH-tus*
plicata, plicatum
Pleated, as in *Thuja plicata*

plumarius *ploo-MAH-ree-us*
plumaria, plumarium
With feathers, as in *Dianthus plumarius*

plumbaginoides *plum-bah-gih-NOY-deez*
Resembling *Plumbago*, as in *Ceratostigma plumbaginoides*

plumbeus *plum-BEY-us*
plumbea, plumbeum
Relating to lead, as in *Alocasia plumbea*

plumosus *plum-OH-sus*
plumosa, plumosum
Feathery, as in *Libocedrus plumosa*

pluriflorus *plur-ee-FLOR-us*
pluriflora, pluriflorum
With many flowers, as in *Erythronium pluriflorum*

pluvialis *ploo-VEE-uh-lis*
pluvialis, pluviale
Relating to rain, as in *Calendula pluvialis*

pocophorus *po-KO-for-us*
pocophora, pocophorum
Fleece-bearing, as in *Rhododendron pocophorum*

podagraria *pod-uh-GRAR-ee-uh*
From *podagra*, Latin for gout, as in *Aegopodium podagraria*

podophyllus *po-do-FIL-us*
podophylla, podophyllum
With stout-stalked leaves, as in *Rodgersia podophylla*

poeticus *po-ET-ih-kus*
poetica, poeticum
Relating to poets, as in *Narcissus poeticus*

PLUMBAGO

The startling blue flowers of *Ceratostigma plumbaginoides*, a plant that provides excellent ground cover.

There are two groups of plants that gardeners frequently refer to as *Plumbago*, which, although both belonging to the *Plumbaginaceae* family, have quite distinct horticultural uses. It is perhaps easiest to distinguish between the two by thinking of one group as primarily containing climbers and the other as consisting of shrubs and ground cover plants. Commonly known as leadwort, the former group of plumbagos are evergreen flowering climbers (and some shrubs) that are not hardy in cool regions; they can often be seen thriving in frost-free conservatories and greenhouses. Their botanical name derives from the Latin for lead, *plumbum*, and first appears in the *Natural History* of Pliny the Elder (23–79 CE), who thought the plant could be used for the treatment of lead poisoning. However, it is also possible that it refers to the intense blue of the flowers common to many of the species. There is also a white form, *Plumbago auriculata* f. *alba*, and a pink one, *P. indica* (syn. *P. rosea*). (*Auriculatus, auriculata, auriculatum* means with an ear or ear-shaped appendage; *indicus, indica, indicum* means of India.)

Known as Chinese or shrubby plumbago, the second group is properly called *Ceratostigma* and consists of half-hardy herbaceous perennial shrubs and ground cover plants. Plants with *cera* in their name derive from the Greek word for horn, *keras*, in this instance referring to the horn-like growth on the stigma of the flower. For ground cover, choose *Ceratostigma plumbaginoides* (meaning it resembles *Plumbago*), which reaches only 30 cm (1 ft) with a spread of 38 cm (15 in). *C. willmottianum* (named after Ellen Willmott, see p. 219) and *C. griffithii* are both very attractive deciduous shrubs that reach upwards of 1 m (3 ft). Although not overtly showy, these plants have some of the loveliest blue flowers one can grow. The blooms are small and subtle yet look particularly good in late autumn sunshine, with the reddening foliage providing an excellent foil. For best results, plant in a warm, sunny position in moist but well-drained soil.

Plumbago indica (syn. *P. rosea*), Scarlet leadwort

Originating from the East Indies, this medium-sized shrub is less commonly grown than the blue-flowered varieties.

polaris *po-LAH-ris*
polaris, polare
Connected with the North Pole, as in *Salix polaris*

polifolius *po-lih-FOH-lee-us*
polifolia, polifolium
With grey leaves, as in *Andromeda polifolia*

politus *POL-ee-tus*
polita, politum
Polished, as in *Saxifraga × polita*

polonicus *pol-ON-ih-kus*
polonica, polonicum
Connected with Poland, as in *Cochlearia polonica*

poly-
Used in compound words to denote many

polyacanthus *pol-lee-KAN-thus*
polyacantha, polyacanthum
With many thorns, as in *Acacia polyacantha*

polyandrus *pol-lee-AND-rus*
polyandra, polyandrum
With many stamens, as in *Conophytum polyandrum*

polyanthemos *pol-ly-AN-them-os*
polyanthus *pol-ee-AN-thus*
polyantha, polyanthum
With many flowers, as in *Jasminum polyanthum*

polyblepharus *pol-ee-BLEF-ar-us*
polyblephara, polyblepharum
With many fringes or eyelashes, as in *Polystichum polyblepharum*

polybotryus *pol-ly-BOT-ree-us*
polybotrya, polybotryum
With many clusters, as in *Acacia polybotrya*

polybulbon *pol-ly-BUL-bun*
With many bulbs, as in *Dinema polybulbon*

polycarpus *pol-ee-KAR-pus*
polycarpa, polycarpum
With many fruits, as in *Fatsia polycarpa*

polycephalus *pol-ee-SEF-a-lus*
polycephala, polycephalum
With many heads, as in *Cordia polycephala*

Veronica polita,
grey field-speedwell

polychromus *pol-ee-KROW-mus*
polychroma, polychromum
With many colours, as in *Euphorbia polychroma*

polygaloides *pol-ee-gal-OY-deez*
Resembling milkwort (*Polygala*), as in *Osteospermum polygaloides*

polygonoides *pol-ee-gon-OY-deez*
Resembling *Polygonum*, as in *Alternanthera polygonoides*

polylepis *pol-ee-LEP-is*
polylepis, polylepe
With many scales, as in *Dryopteris polylepis*

polymorphus *pol-ee-MOR-fus*
polymorpha, polymorphum
With many or variable forms, as in *Acer polymorphum*

polypetalus *pol-ee-PET-uh-lus*
polypetala, polypetalum
With many petals, as in *Caltha polypetala*

polyphyllus *pol-ee-FIL-us*
polyphylla, polyphyllum
With many leaves, as in *Paris polyphylla*

polypodioides *pol-ee-pod-ee-OY-deez*
Resembling *Polypodium*, as in *Blechnum polypodioides*

polyrhizus *pol-ee-RY-zus*
polyrhiza, polyrhizum
polyrrhizus
polyrrhiza, polyrrhizum
With many roots, as in *Allium polyrrhizum*

polysepalus *pol-ee-SEP-a-lus*
polysepala, polysepalum
With many sepals, as in *Nuphar polysepala*

polystachyus *pol-ee-STAK-ee-us*
polystachya, polystachyum
With many spikes, as in *Ixia polystachya*

polystichoides *pol-ee-stik-OY-deez*
Resembling *Polystichum*, as in *Woodsia polystichoides*

polytrichus *pol-ee-TRY-kus*
polytricha, polytrichum
With many hairs, as in *Thymus politrichus*

pomeridianus *pom-er-id-ee-AHN-us*
pomeridiana, pomeridanium
Flowering in the afternoon, as in *Carpanthea pomeridiana*

pomiferus *pom-IH-fer-us*
pomifera, pomiferum
Bearing apples, as in *Maclura pomifera*

pomponius *pomp-OH-nee-us*
pomponia, pomponium
With a tuft or pompon, as in *Lilium pomponium*

ponderosus *pon-der-OH-sus*
ponderosa, ponderosum
Heavy, as in *Pinus ponderosa*

ponticus *PON-tih-kus*
pontica, ponticum
Connected with Pontus, Asia Minor, as in *Daphne pontica*

populifolius *pop-yoo-lih-FOH-lee-us*
populifolia, populifolium
With leaves like poplar (*Populus*), as in *Cistus populifolius*

populneus *pop-ULL-nee-us*
populnea, populneum
Relating to poplar (*Populus*), as in *Brachychiton populneus*

porophyllus *po-ro-FIL-us*
porophylla, porophyllum
With leaves with (apparent) holes, as in *Saxifraga porophylla*

porphyreus *por-FY-ree-us*
porphyrea, porphyreum
Purple-red, as in *Epidendrum porphyreum*

porrifolius *po-ree-FOH-lee-us*
porrifolia, porrifolium
With leaves like leek (*Allium porrum*), as in *Tragopogon porrifolius*

porrigens *por-RIG-enz*
Spreading, as in *Schizanthus porrigens*

portenschlagianus *port-en-shlag-ee-AH-nus*
portenschlagiana, portenschlagianum
Named after Franz von Portenschlag-Leydermayer
(1772–1822), Austrian naturalist, as in *Campanula portenschlagiana*

poscharskyanus *po-shar-skee-AH-nus*
poscharskyana, poscharskyanum
Named after Gustav Poscharsky (1832–1914), German
horticulturist, as in *Campanula poscharskyana*

LATIN IN ACTION

This stunning lily is native to the limestone gorges of
southern France and is sadly becoming increasingly
rare. With its bright red flowers spotted with black,
it grows up to a stately 1 m (3 ft) tall. Some people
find its fragrance unpleasant.

Lilium pomponium, turban lily

Numbers and Plants

Numbers often feature in descriptive botanical terms, where they are used to enumerate the quantity of a particular part of a plant such as the number of its petals, leaves, stamens or colours. You will encounter both the Latin and Greek system of numbers, so it is worth becoming familiar with each, at least from one to ten. The number is always expressed as a prefix to the word, as in the following examples, in which the Latin is given first followed by the Greek equivalent. It is not correct to mix number prefixes derived from Latin with suffixes derived from the Greek, although some terms, such as *-lobus* (lobed), are used with both Latin and Greek.

If describing a plant with only one colour using a Latin prefix, the term is *unicolor*, while one with a single wing using the Greek would be *monopterus*. (Here the Latin for one is *uni-* and the Greek *mono-*.) A biflorus plant has twin flowers and a dispermus one two seeds (two is *bi-* and *di-*). The Latin and Greek for three is the same, *tri-*; for example *tricephalus*, with three heads. *Quadridentatus* is with four teeth, while a plant described as *tetrachromus* would have four colours (four is *quadri-* and *tetra-*). Five stamens is *quinquestamineus* and five angles *pentagonus* (five is *quinque-* and *penta-*). Six-angled is *sexangularis* and six-petalled *hexapetalus* (six is *sex-* and *hexa-*). Seven-lobed would be *septemlobus* and with seven leaves *heptaphyllus* (seven is *septem-* and *hepta-*). Both the Latin and Greek for eight is *octo-*, as in *octosepalus*, with eight sepals. A *novempunctatus* plant would have the distinctive marks of nine spots, while one with nine leaves would be called *enneaphyllus* (nine is *novem-* and *ennea-*). Finally, *decemangulus* is with ten angles and *decapleurus* with ten veins (*decem-* and *deca-*).

Commonly encountered terms that relate to numbers include *biennis* (*biennis, bienne*) meaning biennial, as in *Linum bienne*, the lovely pale flax. *Unicolor* and *bicolor* also occur quite frequently as part of a name, for instance *Fritillaria meleagris* var. *unicolor* subvar. *alba*, the white-flowered snake's head fritillary. Houseplant fans may wish to grow *Dracaena marginata* 'Tricolor', the little dragon tree with variegated striped

Narcissus biflorus,
twin sisters

Biflorus means 'with twin flowers', as
in this unusual and lovely narcissus.

Viola tricolor
wild pansy or heartsease

The heartsease *Viola tricolor* is a good example of a flower
with three distinct colours.

Primula 'Burnard's Formosa', a 19th-century polyanthus

Polyanthus is one of the many horticultural groups into which
the large genus *Primula* has been divided. The name comes from
the Greek *poly-*, many, and *anthos*, flower.

foliage. *Duplex* and *duplicatus* (*duplicata*, *duplicatum*)
both mean double or duplicate. Not surprisingly,
multi- means many, as in *multibracteatus*
(*multibracteata*, *multibracteatum*), with many bracts.
Salvia multicaulis (*multicaulis*, *multicaule*) is known
as the many-stemmed sage. The species epithet of the
American wildflower *Baileya multiradiata* means
with many rays and refers to the ray florets of its
yellow flowers (*multiradiatus*, *multiradiata*, *multiradiatum*). *Myri-* means very many; thus *myriocarpus*
(*myriocarpa*, *myriocarpum*) means with very many
fruits, as in the paddy melon *Cucumis myriocarpus*.
An attractive aquatic plant is *Potamogeton octandrus*,
its name indicating that it has eight stamens
(*octandra*, *octandrum*).

For a flower-filled garden, growers should look for
plants named *myrianthus* (*myriantha*, *myrianthum*),
with many flowers, hence the aptly named *Allium
myrianthum*. This delightful plant has more than a
hundred tiny milky-white blooms on each globe-like
flower head. Another plant name associated with
numbers is *Cotoneaster hebephyllus* var. *monopyrenus*,
the one-stoned cotoneaster. *Hebephyllus* (*hebephylla*,
hebephyllum) means with down-covered leaves;
monopyrenus (*monopyrena*, *monopyrenum*)
means with one stone. The rather unattractively
named scurvy-grass sorrel is properly known as
Oxalis enneaphylla. *Enneaphyllus* (*enneaphylla*,
enneaphyllum) means with nine leaves; in this case,
it refers to the distinctively divided leaves of this
herbaceous perennial alpine.

potamophilus *pot-am-OH-fil-us*
potamophila, potomaphilum
River-loving, as in *Begonia potamophila*

potaninii *po-tan-IN-ee-eye*
Named after Grigory Nikolaevich Potanin (1835–1920), Russian plant collector, as in *Indigofera potaninii*

potatorum *poh-tuh-TOR-um*
Relating to drinking and brewing, as in *Agave potatorum*

---

pottsii *POT-see-eye*
Named after John Potts or C.H. Potts, 19th-century English horticulturists and plant collectors, as in *Crocosmia pottsii*

powellii *pow-EL-ee-eye*
Named after John Wesley Powell (1834–1902), American explorer, as in *Crinum × powellii*

praealtus *pray-AL-tus*
praealta, praealtum
Very tall, as in *Aster praealtus*

praecox *pray-koks*
Very early, as in *Stachyurus praecox*

praemorsus *pray-MOR-sus*
praemorsa, praemorsum
With the appearance of bitten tips, as in *Banksia praemorsa*

praeruptorum *pray-rup-TOR-um*
Growing in rough ground, as in *Peucedanum praeruptorum*

praestans *PRAY-stanz*
Distinguished, as in *Tulipa praestans*

praetextus *pray-TEX-tus*
praetexta, praetextum
With a border, as in *Oncidium praetextum*

prasinus *pra-SEE-nus*
prasina, prasinum
The colour of leeks, as in *Dendrobium prasinum*

pratensis *pray-TEN-sis*
pratensis, pratense
From the meadow, as in *Geranium pratense*

prattii *PRAT-tee-eye*
Named after Antwerp E. Pratt, 19th-century English zoologist, as in *Anemone prattii*

pravissimus *prav-ISS-ih-mus*
pravissima, pravissimum
Very crooked, as in *Acacia pravissima*

primula *PRIM-yew-luh*
First flowering, as in *Rosa primula*

primuliflorus *prim-yoo-LIF-flor-us*
primuliflora, primuliflorum
With flowers like primrose (*Primula*), as in *Rhododendron primuliflorum*

primulifolius *prim-yoo-lih-FOH-lee-us*
primulifolia primulifolium
With leaves like primrose (*Primula*), as in *Campanula primulifolia*

primulinus *prim-yoo-LEE-nus*
primulina, primulinum
primuloides *prim-yoo-LOY-deez*
Like primrose (*Primula*), as in *Paphiopedilum primulinum*

princeps *PRIN-keps*
Most distinguished, as in *Centaurea princeps*

pringlei *PRING-lee-eye*
Named after Cyrus Guernesey Pringle (1838–1911), American botanist and plant collector, as in *Monarda pringlei*

prismaticus *priz-MAT-ih-kus*
prismatica, prismaticum
In the shape of a prism, as in *Rhipsalis prismatica*

proboscideus *pro-bosk-ee-DEE-us*
proboscidea, proboscideum
Shaped like a snout, as in *Arisarum proboscideum*

procerus *PRO-ker-us*
procera, procerum
Tall, as in *Abies procera*

procumbens *pro-KUM-benz*
Prostrate, as in *Gaultheria procumbens*

procurrens *pro-KUR-enz*
Spreading underground, as in *Geranium procurrens*

prodigiosus *pro-dij-ee-OH-sus*
prodigiosa, prodigiosum
Wonderful; enormous; prodigious, as in *Tillandsia prodigiosa*

productus *pro-DUK-tus*
producta, productum
Lengthened, as in *Costus productus*

prolifer *PRO-leef-er*
proliferus *pro-LIH-fer-us*
prolifera, proliferum
Increasing by the production of side shoots, as in *Primula prolifer*

prolificus *pro-LIF-ih-kus*
prolifica, prolificum
Producing many fruits, as in *Echeveria prolifica*

propinquus *prop-IN-kwus*
propinqua, propinquum
Related to, near, as in *Myriophyllum propinquum*

prostratus *prost-RAH-tus*
prostrata, prostratum
Growing flat on the ground, as in *Veronica prostrata*

protistus *pro-TISS-tus*
protista, protistum
The first, as in *Rhododendron protistum*

provincialis *pro-vin-ki-ah-lis*
provincialis, provinciale
Connected with Provence, France, as in *Arenaria provincialis*

pruinatus *proo-in-AH-tus*
pruinata, pruinatum
pruinosus *proo-in-NOH-sus*
pruinosa, pruinosum
Glistening like frost, as in *Cotoneaster pruinosus*

prunelloides *proo-nel-LOY-deez*
Resembling self-heal (*Prunella*), as in *Haplopappus prunelloides*

prunifolius *proo-ni-FOH-lee-us*
prunifolia, prunifolium
With leaves like plum (*Prunus*), as in *Malus prunifolia*

przewalskianus *prez-WAL-skee-ah-nus*
przewalskiana, przewalskianum
przewalskii *prez-WAL-skee*
Named after Nicolai Przewalski, 19th-century Russian naturalist, as in *Ligularia przewalskii*

pseud-
Used in compound words to denote false

pseudacorus *soo-DA-ko-rus*
Deceptively like *Acorus* or sweet flag, as in *Iris pseudacorus*

pseudocamellia *soo-doh-kuh-MEE-lee-uh*
Deceptively like a camellia, as in *Stewartia pseudocamellia*

pseudochrysanthus *soo-doh-kris-AN-thus*
pseudochrysantha, pseudochrysanthum
Resembling a *chrysanthus* species in the same genus, as in *Rhododendron pseudochrysanthum*, which means resembling *R. chrysanthus*

pseudodictamnus *soo-do-dik-TAM-nus*
Deceptively like *Dictamnus*, as in *Ballota pseudodictamnus*

pseudonarcissus *soo-doh-nar-SIS-us*
Deceptively like *Narcissus*; in *N. pseudonarcissus*, it means like *N. poeticus*

psilostemon *sigh-loh-STEE-mon*
With smooth stamens, as in *Geranium psilostemon*

psittacinus *sit-uh-SIGN-us*
psittacina, psittacinum
psittacorum *sit-a-KOR-um*
Like a parrot, relating to parrots, as in *Vriesea psittacinum*

ptarmica *TAR-mik-uh*
ptarmica, ptarmicum
Ancient Greek name for a plant (probably sneezewort) that caused sneezing, as in *Achillea ptarmica*

Hippeastrum psittacinum, amaryllis

pteridoides *ter-id-OY-deez*
Resembling *Pteris*, as in *Coriaria pteridioides*

pteroneurus *ter-OH-new-rus*
pteroneura, pteroneurum
With nerves that have wings, as in *Euphorbia pteroneura*

pubens *PEW-benz*
pubescens *pew-BESS-enz*
Downy, as in *Primula* × *pubescens*

pubigerus *pub-EE-ger-us*
pubigera, pubigerum
Producing down, as in *Schefflera pubigera*

pudicus *pud-IH-kus*
pudica, pudicum
Shy, as in *Mimosa pudica*

pugioniformis *pug-ee-oh-nee-FOR-mis*
pugioniformis, pugioniforme
Shaped like a dagger, as in *Celmisia pugioniformis*

pulchellus *pul-KELL-us*
pulchella, pulchellum
pulcher *PUL-ker*
pulchra, pulchrum
Pretty, beautiful, as in *Correa pulchella*

pulcherrimus *pul-KAIR-ih-mus*
pulcherrima, pulcherrimum
Very beautiful, as in *Dierama pulcherrimum*

pulegioides *pul-eg-ee-OY-deez*
Like *Mentha pulegium* (pennyroyal), as in *Thymus pulegioides*

pulegium *pul-ee-GEE-um*
Latin for pennyroyal, reputed to be a flea-repellent, as in *Mentha pulegium*

pullus *PULL-us*
pulla, pullum
Dark-coloured, as in *Campanula pulla*

pulverulentus *pul-ver-oo-LEN-tus*
pulverulenta, pulverulentum
Appearing to be covered in dust, as in *Primula pulverulenta*

pulvinatus *pul-vin-AH-tus*
pulvinata, pulvinatum
Like a cushion, as in *Echeveria pulvinata*

PULMONARIA

Early herbalists often followed the diagnostic ideas prescribed by the Doctrine of Signatures. This suggested that the physical appearance of a plant offered clues to its beneficial properties for humans. For instance, eyebright (*Euphrasia*) was thought to resemble the human eye and was used to treat optical problems, hence its common name. Similarly, *Dentaria* has tooth-like scales on its rhizomes and was applied to ease the pain of toothache; thus one of its common names is toothwort (*dens* is Latin for tooth).

The distinctive leaves of *Pulmonaria* suggested the shape of the lungs to early herbalists, and the white spots on the foliage were thought to indicate disease, leading them to use preparations made from the plant to treat respiratory problems. *Pulmo* is Latin for lung and the common name is lungwort (the suffix –*wort* alludes to its medicinal use). It has been suggested that *Pulmonaria saccharata* was so named as the white-spotted leaves appear to have been dusted with sugar (*saccharatus*, *saccharata*, *saccharatum* simply means sweet or as if dusted with sugar).

Quite apart from its anatomical associations, *Pulmonaria* has collected a whole range of other common names. Many have biblical connections and include Bethlehem sage, Jerusalem sage, Joseph and Mary and Adam and Eve. Some species bear both blue and pink flowers, and this doubtless inspired the names

On pulmonarias, the appearance of both blue and pink flowers on the same plant has led to common names such as boys and girls.

boys and girls, soldier and his wife, and soldiers and sailors. Rather less poetic is another name, spotted dog.

Pulmonaria is a member of the *Boraginaceae* family, and numerous named varieties have been cultivated. These include several pure-colour forms as well as the bicoloured cultivars. Given favourable growing conditions – namely cool, moist shade – *Pulmonaria* will spread quickly if left unchecked. This makes it an excellent ground cover plant, although it can be a job to eradicate from unwanted areas. Bees love feasting on nectar from the flowers.

Prominent white spots on the leaves are a common feature of many species of this plant.

pumilus *POO-mil-us*
pumila, pumilum
Dwarf, as in *Trollius pumilus*

pumilio *poo-MIL-ee-oh*
Small, dwarf; as in *Edraianthus pumilio*

punctatus *punk-TAH-tus*
punctata, punctatum
With spots, as in *Anthemis punctata*

pungens *PUN-genz*
With a sharp point, as in *Elymus pungens*

puniceus *pun-IK-ee-us*
punicea, puniceum
Red-purple, as in *Clianthus puniceus*

purpurascens *pur-pur-ASS-kenz*
Becoming purple, as in *Bergenia purpurascens*

purpuratus *pur-pur-AH-tus*
purpurata, purpuratum
Made purple, as in *Phyllostachys purpurata*

purpureus *pur-PUR-ee-us*
purpurea, purpureum
Purple, as in *Digitalis purpurea*

purpusii *pur-PUSS-ee-eye*
Named after Carl Purpus (1851–1941) or his brother
Joseph Purpus (1860–1932), German plant collectors,
as in *Lonicera × purpusii*

pusillus *pus-ILL-us*
pusilla, pusillum
Very small, as in *Soldanella pusilla*

pustulatus *pus-tew-LAH-tus*
pustulata, pustulatum
Appearing to be blistered, as in *Lachenalia pustulata*

pycnacanthus *pik-na-KAN-thus*
pycnacantha, pycnacanthum
Densely spined, as in *Coryphantha pycnacantha*

pycnanthus *pik-NAN-thus*
pycnantha, pycnanthum
With densely crowded flowers, as in *Acer pycnanthum*

pygmaeus *pig-MAY-us*
pygmaea, pygmaeum
Dwarf; pygmy, as in *Erigeron pygmaeus*

pyramidalis *peer-uh-mid-AH-lis*
pyramidalis, pyramidale
Shaped like a pyramid, as in *Ornithogalum pyramidale*

pyrenaeus *py-ren-AY-us*
pyrenaea, pyrenaeum
pyrenaicus *py-ren-AY-ih-kus*
pyrenaica, pyrenaicum
Connected with the Pyrenees, as in *Fritillaria pyrenaica*

pyrifolius *py-rih-FOH-lee-us*
pyrifolia, pyrifolium
With leaves like pear (*Pyrus*), as in *Salix pyrifolia*

pyriformis *py-rih-FOR-mis*
pyriformis, pyriforme
Shaped like a pear, as in *Rosa pyriformis*

Q

quadr-
Used in compound words to denote four

quadrangularis *kwad-ran-gew-LAH-ris*
quadrangularis, quadrangulare
quadrangulatus *kwad-ran-gew-LAH-tus*
quadrangulata, quadrangulatum
With four angles, as in *Passiflora quadrangularis*

quadratus *kwad-RAH-tus*
quadrata, quadratum
In fours, as in *Restio quadratus*

quadriauritus *kwad-ree-AWR-ry-tus*
quadriaurita, quadriauritum
With four ears, as in *Pteris quadriaurita*

quadrifidus *kwad-RIF-ee-dus*
quadrifida, quadrifidum
Cut into four, as in *Calothamnus quadrifidus*

quadrifolius *kwod-rih-FOH-lee-us*
quadrifolia, quadrifolium
With four leaves, as in *Marsilea quadrifolia*

quadrivalvis *kwad-rih-VAL-vis*
quadrivalvis, quadrivalve
With four valves, as in *Nicotiana quadrivalvis*

quamash *KWA-mash*
Nex Perce (Native American) word for *Cammasic*, especially
C. quamesh

quamoclit *KWAM-oh-klit*
Old generic name, possibly meaning kidney bean, as in
Ipomoea quamoclit

quercifolius *kwer-se-FOH-lee-us*
quercifolia, quercifolium
With leaves like oak (*Quercus*), as in *Hydrangea quercifolia*

quin-
Used in compound words to denote five

quinatus *kwi-NAH-tus*
quinata, quinatum
In fives, as in *Akebia quinata*

Paris quadrifolia,
herb Paris or true lover's knot

quinoa *KEEN-oh-a*
A Spanish word for *Chenopodium quinoa*, from Quechua, *kinua*

quinqueflorus *kwin-kway-FLOR-rus*
quinqueflora, quinqueflorum
With five flowers, as in *Enkianthus quinqueflorus*

quinquefolius *kwin-kway-FOH-lee-us*
quinquefolia, quinquefolium
With five leaves, often referring to leaflets, as in
Parthenocissus quinquefolia

quinquevulnerus *kwin-kway-VUL-ner-us*
quinquevulnera, quinquevulnerum
With five wounds (i.e., marks), as in *Aerides quinquevulnerum*

QUERCUS

The mighty oak tree features in many myths and legends, and is an enduring symbol of strength.

There appear to be more myths, legends and superstitions surrounding *Quercus*, the oak tree, than most other plants. The tree was associated with the Roman god Jupiter and held sacred by the ancient Druids, who performed their rituals in oak groves. The Norse god Thor is responsible for several superstitions linking the tree with lightning, as he is said to have sheltered beneath an oak tree during a ferocious storm and emerged unscathed. Contrary to sound advice, the superstition of seeking protection from lightning under the oak thus grew (but should not be followed!). Continuing this theme, legend has it that if acorns are collected from a tree that has been struck by lightning, brought home and placed on a windowsill, the house and its inhabitants will be protected from future strikes. By association, the undoubted longevity of the oak is said to impart a long and healthy life to those who carry acorns in their pockets.

The oak's supposed power to protect is reflected in the species name of the English oak, *robur*, which was the Latin word for oak-wood, but also meant strength. Other names hint at horticultural merit: *Quercus rubra* (*ruber, rubra, rubrum* meaning red) and *Quercus coccinea* (*coccineus, coccinea, coccineum* meaning scarlet) are so named in reference to the brilliant colour of their autumn foliage. It might be assumed that the evergreen holm oak, *Quercus ilex*, owes its name to its likeness to holly (*Ilex aquifolium*), but the opposite is true, *ilex* having been the name used for that oak by the Romans.

Several other plants have names associated with the oak, such as *Hydrangea quercifolia*, commonly known as the oakleaf hydrangea (*quercifolius, quercifolia, quercifolium* describes a plant that has leaves like the oak tree). Where *quercinus, quercina* or *quercinum* forms part of a name, it suggests that the plant bears some relationship to the oak; thus *Leccinum quercinum* is a mushroom that is found growing underneath oak trees.

Quercus suber,
Cork oak

The distinctive shapes of the *Quercus* leaf and the acorn have been used as decorative motifs throughout the centuries.

R

racemiflorus *ray-see-mih-FLOR-us*
racemiflora, racemiflorum
racemosus *ray-see-MOH-sus*
racemosa, racemosum
With flowers that appear in racemes, as in *Nepeta racemosa*

raddianus *rad-dee-AH-nus*
raddiana, raddianum
Named after Giuseppe Raddi (1770–1829), Italian botanist,
as in *Adiantum raddianum*

radiatus *rad-ee-AH-tus*
radiata, radiatum
With rays, as in *Pinus radiata*

radicans *RAD-ee-kanz*
With stems that take root, as in *Campsis radicans*

radicatus *rad-ee-KAH-tus*
radicata, radicatum
With conspicuous roots, as in *Papaver radicatum*

radicosus *ray-dee-KOH-sus*
radicosa, radicosum
With many roots, as in *Silene radicosa*

radiosus *ray-dee-OH-sus*
radiosa, radiosum
With many rays, as in *Masdevallia radiosa*

radula *RAD-yoo-luh*
From Latin *radula*, a scraper, as in *Silphium radula*

ramentaceus *ra-men-TA-see-us*
ramentacea, ramentaceum
Covered with scales, as in *Begonia ramentacea*

ramiflorus *ram-ee-FLOR-us*
ramiflora, ramiflorum
With flowers on the older branches, as in *Romulea ramiflora*

ramondioides *ram-on-di-OY-deez*
Resembling *Ramonda*, as in *Conandron ramondoides*

ramosus *ram-OH-sus*
ramosa, ramosum
Branched, as in *Anthericum ramosum*

ramosissimus *ram-oh-SIS-ih-mus*
ramosissima, ramosissimum
Much branched, as in *Lonicera ramosissima*

ramulosus *ram-yoo-LOH-sus*
ramulosa, ramulosum
Twiggy, as in *Celmisia ramulosa*

LATIN IN ACTION

The Latin term *radicans* indicates a plant that has stems that take root readily. One such is *Campsis radicans*; with its trumpet-shaped flowers, it is also known as trumpet creeper and hummingbird vine. Other plants with *radicans* in their name include *Woodwardia radicans*, the rooting chain fern. This plant has huge fronds that can grow up to 2 m (6 ft) long; rather unusually, it develops little rooting plantlets that appear at the tips of their fronds. With its bright orange and red flowers, *Epidendrum radicans* will produce plantlets along its stems and is one of the easiest orchids to grow. Its roots grow all along the length of its stem. Of less use to gardeners is *Rhus radicans*, the invasive and toxic poison ivy, as its sap can cause a very irritating skin rash.

Campsis radicans,
trumpet creeper

ranunculoides *ra-nun-kul-OY-deez*
Resembling buttercup (*Ranunculus*), as in *Anemone ranunculoides*

rariflorus *rar-ee-FLOR-us*
rariflora, rariflorum
With scattered flowers, as in *Carex rariflora*

re-
Used in compound words to denote back or again

reclinatus *rek-lin-AH-tus*
reclinata, reclinatum
Bent backwards, as in *Phoenix reclinata*

rectus *REK-tus*
recta, rectum
Upright, as in *Phygelius × rectus*

recurvatus *rek-er-VAH-tus*
recurvata, recurvatum
recurvus *re-KUR-vus*
recurva, recurvum
Curved backward, as in *Beaucarnea recurvata*

redivivus *re-div-EE-vus*
rediviva, redivivum
Revived; brought back to life (e.g. after drought) as in
Lunaria rediviva

reductus *red-UK-tus*
reducta, reductum
Dwarf, as in *Sorbus reducta*

reflexus *ree-FLEKS-us*
reflexa, reflexum
refractus *ray-FRAK-tus*
refracta, refractum
Bent sharply backwards, as in *Correa reflexa*

refulgens *ref-FUL-genz*
Shining brightly, as in *Bougainvillea refulgens*

regalis *re-GAH-lis*
regalis, regale
Regal; of exceptional merit, as in *Osmunda regalis*

reginae *ree-JIN-ay-ee*
Relating to a queen, as in *Strelitzia reginae*

reginae-olgae *ree-JIN-ay-ee OL-gy*
Named after Queen Olga of Greece (1851–1926), as in
Galanthus reginae-olgae

regius *REE-jee-us*
regia, regium
Royal, as in *Juglans regia*

rehderi *REH-der-eye*
rehderianus *re-der-ee-AH-nus*
rehderiana, rehderianum
Named after Alfred Rehder (1863–1949), German-born
dendrologist who worked at the Arnold Arboretum, Massachusetts,
USA, as in *Clematis rehderiana*

rehmannii *re-MAN-ee-eye*
Named after Joseph Rehmann (1753–1831), German physician, or
Anton Rehmann (1840–1917), Polish botanist, as in *Zantedeschia
rehmannii*

reichardii *ri-KAR-dee-eye*
Named after Johann Jakob Reichard, (1743–1782), German
botanist, as in *Erodium reichardii*

reichenbachiana *rike-en-bak-ee-AH-nuh*
reichenbachii *ry-ken-BAHK-ee-eye*
Named after Heinrich Gottlieb Ludwig Reichenbach (1793–1879 or
Heinrich Gustav Reichenbach, as in *Echinocereus reichenbachii*

religiosus *re-lij-ee-OH-sus*
religiosa, religiosum
Relating to religious ceremonies; sacred, as in *Ficus religiosa*, under
which the Buddha attained enlightenment

remotus *ree-MOH-tus*
remota, remotum
Scattered, as in *Carex remota*

renardii *ren-AR-dee-eye*
Named after Charles Claude Renard (1809–86), as in
Geranium renardii

reniformis *ren-ih-FOR-mis*
reniformis, reniforme
Shaped like a kidney, as in *Begonia reniformis*

repandus *REP-an-dus*
repanda, repandum
With wavy margins, as in *Cyclamen repandum*

repens *REE-penz*
With a creeping habit, as in *Gypsophila repens*

replicatus *rep-lee-KAH-tus*
replicata, replicatum
Doubled; folded back, as in *Berberis replicata*

reptans *REP-tanz*
With a creeping habit, as in *Ajuga reptans*

requienii *re-kwee-EN-ee-eye*
Named after Esprit Requien (1788–1851), French naturalist, as in *Mentha requienii*

resiniferus *res-in-IH-fer-us*
resinifera, resiniferum
resinosus *res-in-OH-sus*
resinosa, resinosum
Producing resin, as in *Euphorbia resinifera*

reticulatus *reh-tick-yoo-LAH-tus*
reticulata, reticulatum
Netted, as in *Iris reticulata*

retortus *re-TOR-tus*
retorta, retortum
retroflexus *ret-roh-FLEKS-us*
retroflexa, retroflexum
retrofractus *re-troh-FRAK-tus*
retrofracta, retrofractum
Twisted or turned backwards, as in *Helichrysum retortum*

retusus *re-TOO-sus*
retusa, retusum
With a rounded and notched tip, *Coryphantha retusa*

reversus *ree-VER-sus*
reversa, reversum
Reversed, as in *Rosa × reversa*

revolutus *re-vo-LOO-tus*
revoluta, revolutum
Rolled backwards (e.g. of leaves), as in *Cycas revoluta*

rex *reks*
King; with outstanding qualities, as in *Begonia rex*

rhamnifolius *ram-nih-FOH-lee-us*
rhamnifolia, rhamnifolium
With leaves like buckthorn (*Rhamnus*), as in *Rubus rhamnifolius*

rhamnoides *ram-NOY-deez*
Resembling buckthorn (*Rhamnus*), as in *Hippophae rhamnoides*

rhizophyllus *ry-zo-FIL-us*
rhizophylla, rhizophyllum
With leaves that take root, as in *Asplenium rhizophyllum*

rhodanthus *rho-DAN-thus*
rhodantha, rhodanthum
With rose-coloured flowers, as in *Mammillaria rhodantha*

rhodopensis *roh-doh-PEN-sis*
rhodopensis, rhodopense
From the Rhodope Mountains, Bulgaria, as in *Haberlea rhodopensis*

rhoeas *RE-as*
Ancient Greek *rhoias*, name for *Papaver rhoeas*

rhombifolius *rom-bih-FOH-lee-us*
rhombifolia, rhombifolium
With diamond-shaped leaves, as in *Cissus rhombifolia*

rhomboideus *rom-BOY-dee-us*
rhomboidea, rhomboideum
Shaped like a diamond, as in *Rhombophyllum rhomboideum*

rhytidophyllus *ry-ti-do-FIL-us*
rhytidophylla, rhytidophyllum
With wrinkled leaves, as in *Viburnum rhytidophyllum*

richardii *rich-AR-dee-eye*
Named after various persons with the forename or surname Richard; thus *Cortaderia richardii* commemorates the French botanist Achille Richard (1794–1852)

Papaver rhoeas,
corn or field poppy

richardsonii *rich-ard-SON-ee-eye*

Named after Sir John Richardson, 19th-century Scottish explorer, as in *Heuchera richardsonii*

rigens *RIG-enz*
rigidus *RIG-ih-dus*
rigida, rigidum

Rigid; inflexible; stiff, as in *Verbena rigida*

rigescens *rig-ES-enz*

Rather rigid, as in *Diascia rigescens*

ringens *RIN-jenz*

Gaping; open, as in *Arisaema ringens*

Geum rivale,
water avens

riparius *rip-AH-ree-us*
riparia, riparium

Of riverbanks, as in *Ageratina riparia*

ritro *RIH-tro*

Probably from the Greek for globe thistle, *rhytros*, as in *Echinops ritro*

ritteri *RIT-ter-ee*
ritterianus *rit-ter-ee-AH-nus*
ritteriana, ritterianum

Named after Friedrich Ritter (1898–1989), German cactus collector, as in *Cleistocactus ritteri*

rivalis *riv-AH-lis*
rivalis, rivale

Growing by the side of streams, as in *Geum rivale*

riversleaianum *riv-ers-lee-i-AY-num*

Named after Riverslea Nursery, Hampshire, England, as in *Geranium × riversleaianum*

riviniana *riv-in-ee-AH-nuh*

Named after Augustus Quirinus Rivinus (August Bachmann; 1652–1723), German physician and botanist, as in *Viola riviniana*

rivularis *riv-yoo-LAH-ris*
rivularis, rivulare

Brook-loving, as in *Cirsium rivulare*

robur *ROH-bur*

Oak, as in *Quercus robur*

robustus *roh-BUS-tus*
robusta, robustum

Growing strongly; sturdy, as in *Eremurus robustus*

rockii *ROK-ee-eye*

Named after Joseph Francis Charles Rock (1884–1962), Austrian-born American plant hunter, as in *Paeonia rockii*

roebelenii *roh-bel-EN-ee-eye*

Named after Carl Roebelen (1855–1927), orchid collector, as in *Phoenix roebelenii*

romanus *roh-MAHN-us*
romana, romanum

Roman, as in *Orchis romana*

romieuxii *rom-YOO-ee-eye*

Named after Henri Auguste Romieux (1857–1937), French botanist, as in *Narcissus romieuxii*

rosa-sinensis *RO-sa sy-NEN-sis*
The rose of China, as in *Hibiscus rosa-sinensis*

rosaceus *ro-ZAY-see-us*
rosacea, rosaceum
Rose-like, as in *Saxifraga rosacea*

roseus *RO-zee-us*
rosea, roseum
Coloured like rose (*Rosa*), as in *Lapageria rosea*

rosmarinifolius *rose-ma-rih-nih-FOH-lee-us*
rosmarinifolia, rosmarinifolium
With leaves like rosemary (*Rosmarinus*), as in
Santolina rosmarinifolia

rostratus *ro-STRAH-tus*
rostrata, rostratum
With a beak, as in *Magnolia rostrata*

rotatus *ro-TAH-tus*
rotata, rotatum
Shaped like a wheel, as in *Phlomis rotata*

rothschildianus *roths-child-ee-AH-nus*
rothschildiana, rothschildianum
Named after Lionel Walter Rothschild (1868–1937), or other members
of the House of Rothschild, as in *Paphiopedilum rothschildianum*

rotundatus *roh-tun-DAH-tus*
rotundata, rotundatum
Rounded, as in *Carex rotundata*

rotundifolius *ro-tun-dih-FOH-lee-us*
rotundifolia, rotundifolium
With leaves that are round, as in *Prostanthera rotundifolia*

rotundus *ro-TUN-dus*
rotunda, rotundum
Rounded, as in *Cyperus rotundus*

rowleyanus *ro-lee-AH-nus*
Named after Gordon Douglas Rowley, (b.1921) British botanist and
succulent expert, as in *Senecio rowleyanus*

roxburghii *roks-BURGH-ee-eye*
Named after William Roxburgh (1751–1815), Superintendent of
Calcutta Botanic Garden, as in *Rosa roxburghii*

roxieanum *rox-ee-AY-num*
Named after Roxie Hanna, 19th-century British missionary,
as in *Rhododendron roxieanum*

rubellus *roo-BELL-us*
rubella, rubellum
Pale red, becoming red, as in *Peperomia rubella*

rubens *ROO-benz*
ruber *ROO-ber*
rubra, rubrum
Red, as in *Plumeria rubra*

rubescens *roo-BES-enz*
Becoming red, as in *Salvia rubescens*

rubiginosus *roo-bij-ih-NOH-sus*
rubiginosa, rubiginosum
Rusty, as in *Ficus rubiginosa*

LATIN IN ACTION

With its round leaves (*rotundifolia* means with
round leaves), the wildflower *Pyrola rotundifolia* is
an uncommon plant for a lightly shaded site, such as
woodland ground cover. A member of the *Ericaceae*
family, its sweetly scented white flowers are borne on
upright stems.

Pyrola rotundifolia,
round-leaved wintergreen

rubioides *roo-bee-OY-deez*
Resembling madder (*Rubia*), as in *Bauera rubioides*

rubri-
Used in compound words to denote red

rubricaulis *roo-bri-KAW-lis*
rubricaulis, rubricaule
With red stems, as in *Actinidia rubricaulis*

rubriflorus *roo-brih-FLOR-us*
rubiflora, rubiflorum
With red flowers, as in *Schisandra rubriflora*

rudis *ROO-dis*
rudis, rude
Coarse, growing on uncultivated ground, as in
Persicaria rudis

rufus *ROO-fus*
rufa, rufum
Red, as in *Prunus rufa*

rufinervis *roo-fi-NER-vis*
rufinervis, rufinerve
With red veins, as in *Acer rufinerve*

rugosus *roo-GOH-sus*
rugosa, rugosum
Wrinkled, as in *Rosa rugosa*

rupestris *rue-PES-tris*
rupestris, rupestre
Of rocky places, as in *Leptospermum rupestre*

rupicola *roo-PIH-koh-luh*
Growing on cliffs and ledges, as in *Penstemon rupicola*

rupifragus *roo-pee-FRAG-us*
rupifraga, rupifragum
Rock-breaking, as in *Papaver rupifragum*

ruscifolius *rus-kih-FOH-lee-us*
ruscifolia, ruscifolium
With leaves like butcher's broom (*Ruscus*), as in
Sarcococca ruscifolia

russatus *russ-AH-tus*
russata, russatum
Russet, as in *Rhododendron russatum*

Acer rufinerve,
snakebark maple

russellianus *russ-el-ee-AH-nus*
russelliana, russellianum
Named after John Russell, 6th Duke of Bedford (1766–1839),
author of numerous botanical and horticultural works, as in
Miltonia russelliana

rusticanus *rus-tik-AH-nus*
rusticana, rusticanum
rusticus *RUS-tih-kus*
rustica, rusticum
Relating to the country, as in *Armoracia rusticana*

ruta-muraria *ROO-tuh-mur-AY-ree-uh*
Literally wall rue, as in *Asplenium ruta-muraria*

ruthenicus *roo-THEN-ih-kus*
ruthenica, ruthenicum
Connected with Ruthenia, a historial area consisting of parts of
Russia and eastern Europe, as in *Fritillaria ruthenica*

rutifolius *roo-tih-FOH-lee-us*
rutifolia, rutifolium
With leaves like rue (*Ruta*), as in *Corydalis rutifolia*

rutilans *ROO-til-lanz*
Reddish, as in *Parodia rutilans*

S

sabatius *sa-BAY-shee-us*
sabatia, sabatium
Connected with Savona, Italy,
as in *Convolvulus sabatius*

saccatus *sak-KAH-tus*
saccata, saccatum
Like a bag, or saccate, as in *Lonicera saccata*

saccharatus *sak-kar-RAH-tus*
saccharata, saccharatum
saccharinus *sak-kar-EYE-nus*
saccharina, saccharinum
Sweet or sugared, as in *Pulmonaria saccharata*

sacciferus *sak-IH-fer-us*
saccifera, sacciferum
Bearing bags or sacks, as in *Dactylorhiza saccifera*

sachalinensis *saw-kaw-lin-YEN-sis*
sachalinensis, sachalinense
From the island Sakhalin, off the coast of Russia, as in *Abies sachalinensis*

sagittalis *saj-ih-TAH-lis*
sagittalis, sagittale
sagittatus *saj-ih-TAH-tus*
sagittata, sagittatum
Shaped like an arrow, as in *Genista sagittalis*

sagittifolius *sag-it-ih-FOH-lee-us*
sagittifolia, sagittifolium
With leaves shaped like arrows, as in *Sagittaria sagittifolia*

salicarius *sa-lih-KAH-ree-us*
salicaria, salicarium
Like willow (*Salix*), as in *Lythrum salicaria*

salicariifolius *sa-lih-kar-ih-FOH-lee-us*
salicariifolia, salicariifolium
salicifolius *sah-lis-ih-FOH-lee-us*
salicifolia, salicifolium
With leaves like willow (*Salix*), as in *Magnolia salicifolia*

salicinus *sah-lih-SEE-nus*
salicina, salicinum
Like willow (*Salix*), as in *Prunus salicina*

salicornioides *sal-eye-korn-ee-OY-deez*
Resembling glasswort (*Salicornia*), as in *Hatiora salicornioides*

salignus *sal-LIG-nus*
saligina, saliginum
Like willow (*Salix*), as in *Podocarpus salignus*

salinus *sal-LY-nus*
salina, salinum
Of salty regions, as in *Carex salina*

The leaves of this aquatic herbaceous perennial plant certainly match their name *sagittifolia* (meaning leaves shaped like an arrow). The edible tubers are considered something of a delicacy in China. They are traditionally consumed at Chinese New Year, and their Chinese name means 'benevolent mushroom'.

Sagittaria sagittifolia,
arrowhead

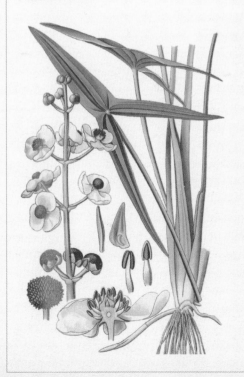

Sir Joseph Hooker

(1817–1911)

Joseph Dalton Hooker was one of the most important British botanists and plant collectors of the 19th century. Through his introduction of Himalayan rhododendrons, he influenced the development of many famous gardens throughout the world. Joseph Hooker was born in Suffolk, but the Hooker family moved north to Scotland when Joseph's father, William Jackson Hooker, was appointed Chair of Botany at Glasgow University. Joseph studied medicine at Glasgow University, but continued to develop his early interest in plants.

In 1839, Joseph Hooker joined the British government's Antarctic expedition, aboard Captain James Clark Ross's ship *Erebus*, as assistant surgeon and botanist. The main task of the four-year expedition was to establish the position of the magnetic South Pole; however, Hooker was also charged with identifying and collecting native plants of economic value. The *Erebus* stopped at places as diverse as Madeira, the Cape of Good Hope, Tasmania, New Zealand, Australia, the Falkland Islands and the southernmost point of South America. Despite many incidents of high drama involving gales and icebergs, the voyage proved to be a great preparation for Hooker's later botanical expeditions.

After returning to England, Hooker set out again in 1847, this time on an expedition to collect plants in India. Initially based at Darjeeling, the intrepid Hooker mounted an elephant and explored the Soane Valley, encountering man-eating tigers and crocodiles along the way. However, it was insects that seemed to have plagued Hooker most and caused the greatest discomfort, not least of all 'a loathesome tick ... as large as the little finger-nail.' He wrote, 'To leeches I am indifferent now ... and other wholesome-looking blood-suckers; but in ticks, as in bugs, there is something revolting to me: the very writing about them makes my flesh creep.'

During the years he spent in India, Hooker travelled widely, including sailing the Ganges River and traversing the mountainous terrain of the Himalayas. His attempt to cross the border and travel in Sikkim caused something of an international incident, as his travelling companion, fellow physician Archibald Campbell, was taken prisoner by locals. Hooker called on the Governor-General Lord Dalhousie for assistance, and Dalhousie sent a

Like many 19th-century botanists, Hooker was a close associate of Charles Darwin and supporter of his evolutionary theories.

regiment of soldiers to the border with plans to invade. Fighting was avoided but the result was that Sikkim was annexed to India, thus becoming part of the British Empire. It was in Campbell's honour that Hooker named *Magnolia campbellii*. In true explorer-of-Empire style, it was not uncommon for Hooker to travel with a team of up to 50 locals, including the all-important guides and guards. He sent a huge number of new plant introductions back to London's Kew Gardens. Among his most important finds were many rhododendrons, including *Rhododendron griffithianum* var. *aucklandii* and *R. edgeworthii*. He named the Himalayan woodland poppy *Cathcartia villosa* after James Ferguson Cathcart, a British civil servant he met in India. Grown today as *Meconopsis villosa*, this plant is still a highly prized addition to many gardens. Cathcart supplied the drawings upon which the artist Walter Hood Fitch was later to base his illustrations for Hooker's 1855 *Illustrations of Himalayan Plants*. Other publications arising from his years in India include the two-volume *Himalayan Journeys*, *Flora Indica* and *The Rhododendrons of Sikkim-Himalaya*.

Back on home turf, Hooker took on the prestigious post of Director of Kew Gardens, a position he held for 20 years. This was to be something of a dynastic role: Hooker's father had previously held the post, and his own son-in-law, Sir William Thiselton-Dyer (who had married Hooker's daughter Harriet) succeeded him. A great friend of Charles Darwin (Hooker was one of those who urged Darwin to publish his ground-breaking work *On the Origin of Species*), Hooker continued

Rhododendron dalhousiae

Collected by Hooker during his time in India, this rhododendron was named for Lady Dalhousie, wife of Lord Dalhousie, Governor-General of India. It has large creamy-white to pale yellow flowers, and Hooker lavished it with praise, describing it as 'the most lovely thing you can imagine' and 'the noblest species of the whole race'.

to travel, visiting the United States in the 1870s and publishing books on the plant life of a wide range of locations. These included *Flora Antarctica*, *Flora Novae-Zelandiae* and *Flora Tasmaniae*. Closer to home, he collaborated with the revered botanist George Bentham (1800–84) on *Genera Plantarum* and the *Handbook of the British Flora*. The latter became something of a bible for students of botany; used for decades, it was known simply as 'Bentham and Hooker'. After a long and impressive career, Hooker died at the great age of 94.

'AT KEW THEIR [THE HOOKERS'] NAMES ARE
REVERED BEYOND ALL OTHERS AND THEIR INFLUENCE
WILL ALWAYS PERVADE THE GARDENS AND IS AN
INSPIRATION TO ALL ASSOCIATED WITH KEW'
Sir George Taylor, Director of Kew Gardens, 1956–1971

saluenensis *sal-WEN-en-sis*
saluenensis, saluenense
From the Salween River, China, as in *Camellia saluenensis*

salviifolius *sal-vee-FOH-lee-us*
salviifolia, salviifolium
With leaves like salvia, as in *Cistus salviifolius*

sambucifolius *sam-boo-kih-FOH-lee-us*
sambucifolia, sambucifolium
With leaves like elder (*Sambucus*), as in *Rodgersia sambucifolia*

sambucinus *sam-byoo-ki-nus*
sambucina, sambucinum
Like elder (*Sambucus*), as in *Rosa sambucina*

samius *SAM-ee-us*
samia, samium
Connected with the isle of Samos, Greece, as in *Phlomis samia*

sanctus *SANK-tus*
sancta, sanctum
Holy, as in *Rhododendron sanctum*

sanderi *SAN-der-eye*
sanderianus *san-der-ee-AH-nus*
sanderiana, sanderianum
Named after Henry Frederick Conrad Sander (1847–1920),
German-born British plant collector, nurseryman and orchid expert,
as in *Dracaena sanderiana*

sanguineus *san-GWIN-ee-us*
sanguinea, sanguineum
Blood-red, as in *Geranium sanguineum*

sapidus *sap-EE-dus*
sapida, sapidum
With a pleasant taste, as in *Rhopalostylis sapida*

saponarius *sap-oh-NAIR-ee-us*
saponaria, saponarium
Soapy, as in *Sapindus saponaria*

sarcocaulis *sar-koh-KAW-lis*
sarcocaulis, sarcocaule
With a fleshy stem, as in *Crassula sarcocaulis*

sarcodes *sark-OH-deez*
Flesh-like, as in *Rhododendron sarcodes*

sardensis *saw-DEN-sis*
sardensis, sardense
Of Sardis (Sart), Turkey, as in *Chionodoxa sardensis*

sargentianus *sar-jen-tee-AH-nus*
sargentiana, sargentianum
sargentii *sar-JEN-tee-eye*
Named after Charles Sprague Sargent (1841–1927),
dendrologist and director of the Arnold Arboretum, Harvard,
USA, as in *Sorbus sargentiana*

sarmaticus *sar-MAT-ih-kus*
sarmatica, sarmaticum
From Sarmatia, historic territory now partly in Poland, partly in
Russia, as in *Campanula sarmatica*

sarmentosus *sar-men-TOH-sus*
sarmentosa, sarmentosum
Producing runners, as in *Androsace sarmentosa*

sarniensis *sarn-ee-EN-sis*
sarniensis, sarniense
From the island of Sarnia (Guernsey), as in *Nerine sarniensis*

sasanqua *suh-SAN-kwuh*
From the Japanese name for *Camellia sasanqua*

sativus *sa-TEE-vus*
sativa, sativum
Cultivated, as in *Castanea sativa*

saundersii *son-DER-see-eye*
Commemorates various eminent Saunders, for example Sir Charles
Saunders (1857–1935); as in *Pachypodium saundersii*

saxatilis *saks-A-til-is*
saxatilis, saxatile
Of rocky places, as in *Aurinia saxatilis*

saxicola *saks-IH-koh-luh*
Growing in rocky places, as in *Juniperus saxicola*

saxifraga *saks-ee-FRAH-gah*
Rock-breaking, as in *Petrorhagia saxifraga*

saxorum *saks-OR-um*
Of the rocks, as in *Streptocarpus saxorum*

saxosus *saks-OH-sus*
saxosa, saxosum
Of rocky places, as in *Gentiana saxosa*

LATIN IN ACTION

Sativus means cultivated; it forms part of the botanical name of several plants with culinary or medicinal uses. *Crocus sativus* is grown for its crimson stigmas; once dried they are sold as the highly prized and expensive spice saffron. Apart from the more notorious uses of *Cannabis sativa* (marijuana), the produce of this annual herb includes a protein-rich birdseed. Rather less contentiously, most vegetable-growers will have enjoyed the pleasure of consuming raw freshly picked and shelled *Pisum sativum*, the garden pea. Indeed, as far back as 7800 BCE, wild strains of peas were gathered and over time selectively improved to become what we recognise today as the garden pea. In the kitchen, few cooks could function without garlic, *Allium sativum*, while *Oryza sativa* is rice.

Pisum sativum,
garden pea

scaber *SKAB-er*
scabra, scabrum
Rough, as in *Eccremocarpus scaber*

scabiosus *skab-ee-OH-sus*
scabiosa, scabiosum
Scabrous, or relating to scabies, as in *Centaurea scabiosa*

scabiosifolius *skab-ee-oh-sih-FOH-lee-us*
scabiosifolia, scabiosifolium
With leaves like scabious (*Scabiosa*), as in *Salvia scabiosifolia*

scalaris *skal-AH-ris*
scalaris, scalare
Like a ladder, as in *Sorbus scalaris*

scandens *SKAN-denz*
Climbing, as in *Cobaea scandens*

scaposus *ska-POH-sus*
scaposa, scaposum
With leafless flowering stems (scapes) as in *Aconitum scaposum*

scariosus *skar-ee-OH-sus*
scariosa, scariosum
Shrivelled, as in *Liatris scariosa*

sceptrum *SEP-trum*
Like a sceptre, as in *Isoplexis sceptrum*

schafta *SHAF-tuh*
A Caspian vernacular name for *Silene schafta*

schidigera *ski-DEE-ger-ruh*
Bearing a spine or splinter, as in *Yucca schidigera*

schillingii *shil-LING-ee-eye*
Named after Tony Schilling (b. 1935), British plantsman, as in *Euphorbia schillingii*

schizopetalus *ski-zo-pe-TAY-lus*
schizopetala, schizopetalum
With cut petals, as in *Hibiscus schizopetalus*

schizophyllus *skits-oh-FIL-us*
schizophylla, schizophyllum
With cut leaves, as in *Syagrus schizophylla*

schmidtianus *shmit-ee-AH-nus*
schmidtiana, schmidtianum
schmidtii *SHMIT-ee-eye*
Commemorates various eminent botanists called Schmidt, as in
Artemisia schmidtiana

schoenoprasum *skee-no-PRAY-zum*
Epithet for chives (*Allium schoenoprasum*), meaning 'rush-leek'
in Greek

schottii *SHOT-ee-eye*
Can commemorate various naturalists called Schott, for example
Arthur Carl Victor Schott (1814–75), as in *Yucca schottii*

schubertii *shoo-BER-tee-eye*
Named after Gotthilf von Schubert (1780–1860),
German naturalist, as in *Allium schubertii*

schumannii *shoo-MAHN-ee-eye*
Named after Dr Karl Moritz Schumann (1851–1904),
German botanist, as in *Abelia schumannii*

scillaris *sil-AHR-is*
scillaris, scillare
Like *Scilla*, as in *Ixia scillaris*

Allium scorodoprasum,
sand leek

scillifolius *sil-ih-FOH-lee-us*
scillifolia, scillifolium
With leaves like *Scilla*, as in *Roscoea scillifolia*

scilloides *sil-OY-deez*
Resembling *Scilla*, as in *Puschkinia scilloides*

scilloniensis *sil-oh-nee-EN-sis*
scilloniensis, scilloniense
From the Isles of Scilly, England, as in *Olearia × scilloniensis*

sclarea *SKLAR-ee-uh*
From *clarus*, clear, as in *Salvia sclarea*

sclerophyllus *skler-oh-FIL-us*
sclerophylla, sclerophyllum
With hard leaves, as in *Castanopsis sclerophylla*

scolopendrius *skol-oh-PEND-ree-us*
scolopendria, scolopendrium
From the Greek word for *Asplenium scolopendrium*, from a supposed
likeness of the underside of its fronds to a millipede or centipede
(Greek *skolopendra*)

scolymus *SKOL-ih-mus*
From the Greek for an edible kind of thistle or artichoke, as in
Cynara scolymus

scoparius *sko-PAIR-ee-us*
scoparia, scoparium
Like broom, as in *Cytisus scoparius*

scopulorum *sko-puh-LOR-um*
Of crags or cliffs, as in *Cirsium scopulorum*

scorodoprasum *skor-oh-doh-PRAY-zum*
Garlic name for a plant between leek and garlic, as in
Allium scorodoprasum

scorpioides *skor-pee-OY-deez*
Resembling a scorpion's tail, as in *Myosotis scorpioides*

scorzonerifolius *skor-zon-er-ih-FOH-lee-us*
scorzonerifolia, scorzonerifolium
With leaves like *Scorzonera*, as in *Allium scorzonerifolium*

scoticus *SKOT-ih-kus*
scotica, scoticum
Connected with Scotland, as in *Primula scotica*

scouleri *SKOOL-er-ee*
Named after Dr John Scouler (1804–71), Scottish botanist, as in *Hypericum scouleri*

scutatus *skut-AH-tus*
scutata, scutatum
scutellaris *skew-tel-AH-ris*
scutellaris, scutellare
scutellatus *skew-tel-LAH-tus*
scutellata, scutellatum
Shaped like a shield or platter, as in *Rumex scutatus*

secundatus *see-kun-DAH-tus*
secundata, secundatum
secundiflorus *sek-und-ee-FLOR-us*
secundiflora, secundiflorum
secundus *se-KUN-dus*
secunda, secundum
With leaves or flowers growing on one side of a stalk only, as in *Echeveria secunda*

seemannianus *see-mahn-ee-AH-nus*
seemanniana, seemannianum
seemannii *see-MAN-ee-eye*
Named after Berthold Carl Seemann (1825–71), German plant collector, as in *Hydrangea seemannii*

segetalis *seg-UH-ta-lis*
segetalis, segetale
segetum *seg-EE-tum*
Of cornfields, as in *Euphorbia segetalis*

selaginoides *sel-ag-ee-NOY-deez*
Resembling clubmoss (*Selaginella*), as in *Athrotaxis selaginoides*

selloanus *sel-lo-AH-nus*
selloana, selloanum
Named after Friedrich Sellow (Sello), 19th-century German explorer and plant collector, as in *Cortaderia selloana*

semperflorens *sem-per-FLOR-enz*
Ever-blooming, as in *Grevillea × semperflorens*

sempervirens *sem-per-VY-renz*
Evergreen, as in *Lonicera sempervirens*

sempervivoides *sem-per-vi-VOY-deez*
Resembling house leek (*Sempervivum*), as in *Androsace sempervivoides*

senegalensis *sen-eh-gal-EN-sis*
senegalensis, senegalense
From Senegal, Africa, as in *Persicaria senegalensis*

senescens *sen-ESS-enz*
Seeming to grow old (i.e. white or grey), as in *Allium senescens*

senilis *SEE-nil-is*
senilis, senile
With white hair, as in *Rebutia senilis*

sensibilis *sen-si-BIL-is*
sensibilis, sensibile
sensitivus *sen-si-TEE-vus*
sensitiva, sensitivum
Sensitive to light or touch, as in *Onoclea sensibilis*

sepium *SEP-ee-um*
Growing along hedgerows, as in *Calystegia sepium*

SEMPERVIVUM

Belonging to the *Crassulaceae* family, the name of the long-lived *Sempervivum* comes from the Latin *semper,* always, and *vivus*, alive. It is hardly surprising then, to find the name 'live forever' among their various common appellations – although rather less obvious is hens and chickens, which is used for various sempervivums (the 'hen' is the main plant, while the 'chickens' are the offsets).

The most commonly used name for *Sempervivum* is houseleek, which is particularly applied to *Sempervivum tectorum. Tectorum* means of the roofs of houses – covering the roof tiles of a dwelling with *S. tectorum* is said to protect against lightning strikes. However, if the said plants are picked from the roof by a stranger this can bring terrible luck, even death. Such associations go back to ancient times and are linked with the Norse god of thunder, Thor, as well as with the Roman god Jupiter. Hence the names Jupiter's beard, Jupiter's eye and the German name Donnersbart (thunder beard). In more recent times, claims have been made for the anti-inflammatory properties of *Sempervivum*. As well as being applied to soothe stings, they have been used as a cure for everything from insomnia to poor eyesight. As its name suggests, *S. arachnoideum* is the cobweb houseleek (*arachnoideus, arachnoidea, arachnoideum* means like a spider's web).

The spider's web-like properties of *Sempervivum arachnoideum.*

A great many species and numerous named cultivars belong to this genus, and include hardy and half-hardy evergreen succulents. Their star-like flowers are borne in sprays that rise from tight mats of shiny foliage. They thrive in dry places such as rock gardens, cracks in walls, or tucked between roof tiles, and will spread quite quickly if happily situated in a sunny spot. For best results, try adding grit or coarse sand if the drainage needs improving.

The erect tight red flowers of *Sempervivum tectorum*, the common houseleek.

sept-
Used in compound words to denote seven

septemfidus *sep-TEM-fee-dus*
septemfida, septemfidum
With seven divisions, as in *Gentiana septemfida*

septemlobus *sep-tem-LOH-bus*
septemloba, septemlobum
With seven lobes, as in *Primula septemloba*

septentrionalis *sep-ten-tree-oh-NAH-lis*
septentrionalis, septentrionale
From the north, as in *Beschorneria septentrionalis*

sericanthus *ser-ee-KAN-thus*
sericantha, sericanthum
With silky flowers, as in *Philadelphus sericanthus*

sericeus *ser-IK-ee-us*
sericea, sericeum
Silky, as *Rosa sericea*

serotinus *se-roh-TEE-nus*
serotina, serotinum
With flowers or fruit late in the season, as in *Iris serotina*

serpens *SUR-penz*
Creeping, as in *Agapetes serpens*

serpyllifolius *ser-pil-ly-FOH-lee-us*
serpyllifolia, serpyllifolium
With leaves like wild or creeping thyme (*Thymus serpyllum*),
as in *Arenaria serpyllifolia*

serpyllum *ser-PIE-lum*
From the Greek word for a kind of thyme, as in *Thymus serpyllum*

serratifolius *sair-rat-ih-FOH-lee-us*
serratifolia, serratifolium
With leaves that are serrated or saw-toothed, as in
Photinia serratifolia

serratus *sair-AH-tus*
serrata, serratum
With small-toothed leaf margins, as in *Zelkova serrata*

serrulatus *ser-yoo-LAH-tus*
serrulata, serrulatum,
With small serrations at the leaf margins, as in
Enkianthus serrulatus

Angraecum sesquipedale,
star of Bethlehem orchid or Darwin's orchid

sesquipedalis *ses-kwee-ped-AH-lis*
sesquipedalis, sesquipedale
Eighteen inches long, as in *Angraecum sesquipedale*

sessili-
Used in compound words to denote stalkless

sessiliflorus *sess-il-ee-FLOR-us*
sessiliflora, sessililforum
With stalkless flowers, as in *Libertia sessiliflora*

sessilifolius *ses-ee-lee-FOH-lee-us*
sessilifolia, sessilifolium
With stalkless leaves, as in *Uvularia sessilifolia*

sessilis *SES-sil-is*
sessilis, sessile
Without a stalk, as in *Trillium sessile*

setaceus *se-TAY-see-us*
setacea, setaceum
With bristles, as in *Pennisetum setaceum*

setchuenensis *sech-yoo-en-EN-sis*
setchuenensis, setchuenense
From Sichuan province, China, as in
Deutzia setchuenensis

seti-
Used in compound words to denote bristled

setiferus *set-IH-fer-us*
setifera, setiferum
With bristles, as in *Polystichum setiferum*

setifolius *set-ee-FOH-lee-us*
setifolia, setifolium
With bristly leaves, as in *Lathyrus setifolius*

setiger *set-EE-ger*
setigerus *set-EE-ger-us*
setigera, setigerum
Bearing bristles, as in *Gentiana setigera*

setispinus *set-i-SPIN-us*
setispina, setispinum
With bristly spines, as in *Thelocactus setispinus*

setosus *set-OH-sus*
setosa, setosum
With many bristles, as in *Iris setosa*

setulosus *set-yoo-LOH-sus*
setulosa, setulosum
With many small bristles, as in *Salvia setulosa*

sex-
Used in compound words to denote six

sexangularis *seks-an-gew-LAH-ris*
sexangularis, sexangulare
With six angles, as in *Sedum sexangulare*

sexstylosus *seks-sty-LOH-sus*
sexstylosa, sexstylosum
With six styles, as in *Hoheria sexstylosa*

sherriffii *sher-RIF-ee-eye*
Named after George Sherriff (1898–1967), Scottish plant
collector, as in *Rhododendron sherriffii*

shirasawanus *shir-ah-sa-WAH-nus*
shirasawana, shirasawanum
Named after Homi (or Miho) Shirasawa (1868–1947),
Japanese botanist, as in *Acer shirasawanum*

sibiricus *sy-BEER-ih-kus*
sibirica, sibiricum
Connected with Siberia, as in *Iris sibirica*

sichuanensis *sy-CHOW-en-sis*
sichuanensis, sichuanense
From Sichuan province, China, as in *Cotoneaster sichuanensis*

siculus *SIK-yoo-lus*
sicula, siculum
From Sicily, Italy, as in *Nectaroscordum siculum*

sideroxylon *sy-der-oh-ZY-lon*
Wood like iron, as in *Eucalyptus sideroxylon*

sieberi *sy-BER-ee*
Named after Franz Sieber (1789–1844), Prague-born botanist
and plant collector, as in *Crocus sieberi*

LATIN IN ACTION

The species name of this pink climber, *Rosa setigera*,
refers to bristly prickles that are scattered all along its
stems. A native of Missouri, USA, and grown as either
a spreading shrub or climber, it has simple, single
blooms that resemble those of *Rosa canina*.

Rosa setigera,
climbing prairie rose

sieboldianus *see-bold-ee-AH-nus*
sieboldiana, sieboldianum
sieboldii *see-bold-ee-eye*
Named after Philipp von Siebold (1796–1866), German doctor
who collected plants in Japan, as in *Magnolia sieboldii*

signatus *sig-NAH-tus*
signata, signatum
Well-marked, as in *Saxifraga signata*

sikkimensis *sik-im-EN-sis*
sikkimensis, sikkimense
From Sikkim, India, as in *Euphorbia sikkimensis*

siliceus *sil-ee-SE-us*
silicea, siliceum
Growing in sand, as in *Astragalus siliceus*

siliquastrum *sil-ee-KWAS-trum*
Roman name for a plant with pods, as in *Cercis siliquastrum*

silvaticus *sil-VAT-ih-kus*
silvatica, silvaticum
silvestris *sil-VES-tris*
silvestris, silvestre
Growing in woodlands, as in *Polystichum silvaticum*

similis *SIM-il-is*
similis, simile
Similar; like, as in *Lonicera similis*

simplex *SIM-plecks*
Simple; without branches, as in *Actaea simplex*

simplicifolius *sim-plik-ih-FOH-lee-us*
simplicifolia, simplicifolium
With simple leaves, as in *Astilbe simplicifolia*

simulans *sim-YOO-lanz*
Resembling, as in *Calochortus simulans*

sinensis *sy-NEN-sis*
sinensis, sinense
From China, as in *Corylopsis sinensis*

sinicus *SIN-ih-kus*
sinica, sinicum
Connected with China, as in *Amelanchier sinica*

Wisteria sinensis,
Chinese wisteria

sinuatus *sin-yoo-AH-tus*
sinuata, sinuatum
With a wavy margin, as in *Salpiglossis sinuata*

siphiliticus *sigh-fy-LY-tih-kus*
siphilitica, siphiliticum
Connected with syphilis, as in *Lobelia siphilitica*

sitchensis *sit-KEN-sis*
sitchensis, sitchense
From Sitka, Alaska, as in *Sorbus sitchensis*

skinneri *SKIN-ner-ee*
Named after George Ure Skinner (1804–67), Scottish plant
collector, as in *Cattleya skinneri*

smilacinus *smil-las-SY-nus*
smilacina, smilacinum
Relating to greenbriar (*Smilax*), as in *Disporum smilacinum*

smithianus *SMITH-ee-ah-nus*
smithiana, smithianum
smithii *SMITH-ee-eye*
May commemorate any of several Smiths, including Sir James
Edward Smith (1759–1828), as in *Senecio smithii*

soboliferus *soh-boh-LIH-fer-us*
sobolifera, soboliferum
With creeping rooting stems, as in *Geranium soboliferum*

socialis *so-KEE-ah-lis*
socialis, sociale
Forming colonies, as in *Crassula socialis*

solidus *SOL-id-us*
solida, solidum
Solid; dense, as in *Corydalis solida*

somaliensis *soh-mal-ee-EN-sis*
somaliensis, somaliense
From Somalia, Africa, as in *Cyanotis somaliensis*

somniferus *som-NIH-fer-us*
somnifera, somniferum
Inducing sleep, as in *Papaver somniferum*

sonchifolius *son-chi-FOH-lee-us*
sonchifolia, sonchifolium
With leaves like sowthistle (*Sonchus*), as in *Francoa sonchifolia*

sorbifolius *sor-bih-FOH-lee-us*
sorbifolia, sorbifolium
With leaves like mountain ash (*Sorbus*), as in *Xanthoceras sorbifolium*

sordidus *SOR-deh-dus*
sordida, sordidum
Dirty-looking, as in *Salix × sordida*

soulangeanus *soo-lan-jee-AH-nus*
soulangeana, soulangeanum
Commemorates Étienne Soulange-Bodin (1774–1846),
French diplomat and secretary to the Société Royale et Centrale
d'Agriculture (now the Académie d'Agriculture de France), who
raised *Magnolia × soulangeana*

spachianus *spak-ee-AH-nus*
spachiana, spachianum
Named after Édouard Spach (1801–79), French botanist,
as in *Genista × spachiana*

sparsiflorus *spar-see-FLOR-us*
sparsiflora, sparsiflorum
With sparse or scattered flowers, as in *Lupinus sparsiflorus*

spathaceus *spath-ay-SEE-us*
spathacea, spathaceum
With a spathe, spathe-like, as in *Salvia spathacea*

spathulatus *spath-yoo-LAH-tus*
spathulata, spathulatum
Spatulate, with a broader, flattened end, as in *Aeonium spathulatum*

speciosus *spee-see-OH-sus*
speciosa, speciosum
Showy, as in *Ribes speciosum*

spectabilis *speck-TAH-bih-lis*
spectabilis, spectabile
Spectacular; showy, as in *Sedum spectabile*

sphaericus *SFAY-rih-kus*
sphaerica, sphaericum
Shaped like a sphere, as in *Mammillaria sphaericus*

sphaerocarpos *sfay-ro-KAR-pus*
sphaerocarpa, sphaerocarpum
With round fruits, as in *Medicago sphaerocarpos*

sphaerocephalon *sfay-ro-SEF-uh-lon*
sphaerocephalus *sfay-ro-SEF-uh-lus*
sphaerocephala, sphaerocephalum
With a round head, as in *Allium sphaerocephalon*

spicant *SPIK-ant*
Word of uncertain origin; possibly a German corruption of *spica*,
spike, tuft, as in *Blechnum spicant*

spicatus *spi-KAH-tus*
spicata, spicatum
With ears that grow in spikes, as in *Mentha spicata*

Papaver somniferum (double form),
opium poppy

spiciformis *spik-ee-FOR-mis*
spiciformis, spiciforme
In the shape of a spike, as in *Celastrus spiciformis*

spicigerus *spik-EE-ger-us*
spicigera, spicigerum
Bearing spikes, as in *Justicia spicigera*

spiculifolius *spik-yoo-lih-FOH-lee-us*
spiculifolia, spiculifolium
Like small spikes, as in *Erica spiculifolia*

spinescens *spy-NES-enz*
spinifex *SPIN-ee-feks*
spinosus *spy-NOH-sus*
spinosa, spinosum
With spines, as in *Acanthus spinosus*

spinosissimus *spin-oh-SIS-ih-mus*
spinosissima, spinosissimum
Very spiny, as in *Rosa spinosissima*

spinulosus *spin-yoo-LOH-sus*
spinulosa, spinulosum
With small spines, as in *Woodwardia spinulosa*

spiralis *spir-AH-lis*
spiralis, spirale
Spiral, as in *Macrozamia spiralis*

splendens *SPLEN-denz*
splendidus *splen-DEE-dus*
splendida, splendidum
Splendid, as in *Fuchsia splendens*

sprengeri *SPRENG-er-ee*
Named after Carl Ludwig Sprenger (1846–1917), German botanist
and plantsman, who bred and introduced many new plants, as in
Tulipa sprengeri

spurius *SPEW-eee-us*
spuria, spurium
False; spurious, as in *Iris spuria*

squalidus *SKWA-lee-dus*
squalida, squalidum
Dirty-looking, dingy, as in *Leptinella squalida*

squamatus *SKWA-ma-tus*
squamata, squamatum
With small scale-like leaves or bracts, as in *Juniperus squamata*

Rosa spinosissima
var. *luteola*,
burnet rose

squamosus *skwa-MOH-sus*
squamosa, squamosum
With many scales, as in *Annona squamosa*

squarrosus *skwa-ROH-sus*
squarrosa, squarrosum
With spreading or curving parts at the extremities, as in
Dicksonia squarrosa

stachyoides *stah-kee-OY-deez*
Resembling betony (*Stachys*), as in *Buddleja stachyoides*

stamineus *stam-IN-ee-us*
staminea, stamineum
With pronounced stamens, as in *Vaccinium stamineum*

standishii *stan-DEE-shee-eye*
Named after John Standish (1814–1875), English nurseryman, who
raised plants collected by Robert Fortune, as in *Lonicera standishii*

stans *stanz*
Erect; upright, as in *Clematis stans*

stapeliiformis *sta-pel-ee-ih-FOR-mis*
stapeliiformis, stapeliiforme
Like *Stapelia*, as in *Ceropegia stapeliiformis*

stellaris *stell-AH-ris*
stellaris, stellare
stellatus *stell-AH-tus*
stellata, stellatum
Starry, as in *Magnolia stellata*

steno-
Used in compound words to denote narrow

stenocarpus *sten-oh-KAR-pus*
stenocarpa, stenocarpum
With narrow fruits, as in *Carex stenocarpa*

stenopetalus *sten-oh-PET-al-lus*
stenopetala, stenopetalum
With narrow petals, as in *Genista stenopetala*

stenophyllus *sten-oh-FIL-us*
stenophylla, stenophyllum
With narrow leaves, as in *Berberis* × *stenophylla*

stenostachyus *sten-oh-STAK-ee-us*
stenostachya, stenostachyum
With narrow spikes, as in *Buddleja stenostachya*

Tulipa clusiana var. *stellata*

sterilis *STER-ee-lis*
sterilis, sterile
Infertile; sterile, as in *Potentilla sterilis*

sternianus *stern-ee-AH-nus*
sterniana, sternianum
sternii *STERN-ee-eye*
Named after Sir Frederick Claude Stern (1884–1967), English
horticulturist and author with a particular interest in gardening on
chalk, as in *Cotoneaster sternianus*

stipulaceus *stip-yoo-LAY-see-us*
stipulacea, stipulaceum
stipularis *stip-yoo-LAH-ris*
stipularis, stipulare
stipulatus *stip-yoo-LAH-tus*
stipulata, stipulatum
With stipules, as in *Oxalis stipularis*

stoechas *STOW-kas*
From *stoichas*, meaning in rows, the Greek name for
Lavandula stoechas

stoloniferus *sto-lon-IH-fer-us*
stolonifera, stoloniferum
With runners that take root, as in *Saxifraga stolonifera*

strepto-
Used in compound words to denote twisted

streptophyllus *strep-toh-FIL-us*
streptophylla, streptophyllum
With twisted leaves, as in *Ruscus streptophyllum*

striatus *stree-AH-tus*
striata, striatum
With stripes, as in *Bletilla striata*

strictus *STRIK-tus*
stricta, strictum
Erect; upright, as in *Penstemon strictus*

strigosus *strig-OH-sus*
strigosa, strigosum
With stiff bristles, as in *Rubus strigosus*

striolatus *stree-oh-LAH-tus*
striolata, striolatum
With fine stripes or lines, as in *Dendrobium striolatum*

STREPTOCARPUS

The prefix *strepto-* is used in compound words to denote twisted; thus we have the term *streptopetalus* (*streptopetala*, *streptopetalum*) to describe a plant that has twisted petals. Similarly, *streptophyllus* (*streptophylla*, *streptophyllum*) means with twisted leaves while *streptosepalus* (*streptosepala*, *streptosepalum*) means with twisted sepals. The lovely freely flowering plant known commonly as Cape primrose has seed capsules that are twisted into a spiral, and this feature has given it its genus name, *Streptocarpus*. *Streptocarpus* literally means having twisted fruit, coming from the Greek for twisted, *streptos*, and for fruit, *karpos*.

Belonging to the family *Gesneriaceae* and originating from southern and eastern Africa, the natural habit of *Streptocarpus* is damp woodland. These plants favour bright conditions but not full sun. In cooler regions, these perennials are tender and should be grown as houseplants or in a frost-free greenhouse. Some species have a rosette of leaves from which the flower stems emerge, whereas others, such as *Streptocarpus dunnii* and *S. wendlandii*, are rather curious in only producing a single large leaf that continues to grow for the entirety of the plant's life. A Mr. E. Dunn from Cape Town, South Africa, discovered the former species in the Transvaal in the late 19th century, and it is named in his honour. *Wendlandii* is named after members of the famous German family of botanists who became successive

This group of plants produces an abundance of flowers over a long period and is available in a wide range of colours.

An example of a streptocarpus that produces only a single leaf.

curators of the Herrenhausen Gardens, in Hanover. Other species include *S. cyaneus*, named after its blue flowers; *S. floribunda*, describing the free-flowering nature of these plants; and *S. silvaticus*, alluding to its native woodland habitat. (*Cyaneus*, *cyanea*, and *cyaneum* meaning blue; *floribundus*, *floribunda*, and *floribundum* meaning very free flowering; *silvaticus*, *silvatica*, and *silvaticum* meaning growing in woods.)

Other plants with the prefix *strepto-* include *Streptopus*. From the family *Liliaceae*, this berry-bearing plant has various common names, including clasping twisted stalk, claspleaf twisted stalk and white twisted stalk (*pous* is the Greek for foot). The marmalade bush, *Streptosolen*, gets its name from the twisted tube of its corolla (*solen* means a tube).

strobiliferus *stroh-bil-IH-fer-us*
strobilifera, strobiliferum
Producing cones, as in *Epidendrum strobiliferum*

strobus *STROH-bus*
From Greek *strobos*, a whirling motion (cf. Greek *strobilos*, pine cone), or Latin *strobus*, an incense-bearing tree in Pliny, as in *Pinus strobus*

strumosus *stroo-MOH-sus*
strumosa, strumosum
With cushion-like swellings. as in *Nemesia strumosa*

struthiopteris *struth-ee-OP-ter-is*
Like an ostrich wing, as in *Matteuccia struthiopteris*

stygianus *sty-jee-AH-nuh*
stygiana, stygianum
Dark, as in *Euphorbia stygiana*

LATIN IN ACTION

The epithet of *Pinus strobus* refers to the large pine cones borne by this sizeable evergreen tree. Native to northeastern USA, its other common names include the northern white pine or soft pine, while in Britain it is sometimes called the Weymouth pine.

Pinus strobus,
eastern white pine

---

stylosus *sty-LOH-sus*
stylosa, stylosum
With pronounced styles, as in *Rosa stylosa*

styracifluus *sty-rak-IF-lu-us*
styraciflua, styracifluum
Producing gum, from *styrax*, the Greek name for storax, as in *Liquidambar styraciflua*

suaveolens *swah-vee-OH-lenz*
With a sweet fragrance, as in *Brugmansia suaveolens*

suavis *SWAH-vis*
suavis, suave
Sweet; with a sweet scent, as in *Asperula suavis*

sub-
Used in compound words to denote a variety of meanings such as almost, partially, slightly, rather, under.

subacaulis *sub-a-KAW-lis*
subacaulis, subacaule
Without much stem, as in *Dianthus subacaulis*

subalpinus *sub-al-PY-nus*
subalpina, subalpinum
Growing at the lower levels of mountain ranges, as in *Viburnum subalpinum*

subcaulescens *sub-kawl-ESS-enz*
With a small stem, as in *Geranium subcaulescens*

subcordatus *sub-kor-DAH-tus*
subcordata, subcordatum
Shaped rather like a heart, as in *Alnus subcordata*

suberosus *sub-er-OH-sus*
suberosa, suberosum
With cork bark, as in *Scorzonera suberosa*

subhirtellus *sub-hir-TELL-us*
subhirtella, subhirtellum
Rather hairy, as in *Prunus × subhirtella*

submersus *sub-MER-sus*
submersa, submersum
Submerged, as in *Ceratophyllum submersum*

subsessilis *sub-SES-sil-is*
subsessilis, subsessile
Fixed, as in *Nepeta subsessilis*

subterraneus *sub-ter-RAY-nee-us*
subterranea, subterraneum
Underground, as in *Parodia subterranea*

subtomentosus *sub-toh-men-TOH-sus*
subtomentosa, subtomentosum
Almost hairy, as in *Rudbeckia subtomentosa*

subulatus *sub-yoo-LAH-tus*
subulata, subulatum
Awl- or needle-shaped, as in *Phlox subulata*

subvillosus *sub-vil-OH-sus*
subvillosa, subvillosum
With rather soft hairs, as in *Begonia subvillosa*

succulentus *suk-yoo-LEN-tus*
succulenta, succulentum
Fleshy; juicy, as in *Oxalis succulenta*

suffrutescens *suf-roo-TESS-enz*
suffruticosus *suf-roo-tee-KOH-sus*
suffruticosa, suffruticosum
Rather shrubby, as in *Paeonia suffruticosa*

sulcatus *sul-KAH-tus*
sulcata, sulcatum
With furrows, as in *Rubus sulcatus*

sulphureus *sul-FER-ee-us*
sulphurea, sulphureum
Sulphur-yellow, as in *Lilium sulphureum*

suntensis *sun-TEN-sis*
suntensis, suntense
Named after Sunte House, Sussex, England, as in *Abutilon × suntense*

superbiens *soo-PER-bee-enz*
superbus *soo-PER-bus*
superba, superbum
Superb, as in *Salvia × superba*

supinus *sup-EE-nus*
supina, supinum
Prostrate, as in *Verbena supina*

surculosus *sur-ku-LOH-sus*
surculosa, surculosum
Producing suckers, as in *Dracaena surculosa*

Paeonia suffruticosa,
tree peony or moutan

suspensus *sus-PEN-sus*
suspensa, suspensum
Hanging, as in *Forsythia suspensa*

sutchuenensis *sech-yoo-en-EN-sis*
sutchuenensis, sutchuenense
From Sichuan province, China, as in *Adonis sutchuenensis*

sutherlandii *suth-er-LAN-dee-eye*
Named after Dr Peter Sutherland (1822–1900), who
discovered *Begonia sutherlandii*

sylvaticus *sil-VAT-ih-kus*
sylvatica, sylvaticum
sylvester *sil-VESS-ter*
sylvestris *sil-VESS-tris*
sylvestris, sylvestre
sylvicola *sil-VIH-koh-luh*
Growing in woodlands, as in *Pinus sylvestris*,
Nyssa sylvatica

syriacus *seer-ee-AH-kus*
syriaca, syriacum
Connected with Syria, as in *Asclepias syriaca*

szechuanicus *se-CHWAN-ih-kus*
szechuanica, szechuanicum
Connected with Szechuan, China, as in
Populus szechuanica

Plants and Animals

If asked to name a plant with animal associations, the first one likely to spring to mind is the dog rose, *Rosa canina*, the wild rose so often found scrambling through country hedgerows. (*Caninus, canina, caninum*, meaning relating to dogs, and therefore inferior.) On the other hand, cat lovers may volunteer *Nepeta cataria*, whose common names include catmint, catnip and catswort (*cataria*, pertaining to cats). These, however, are just a small sample of the many plant names associated with the animal kingdom; indeed, there seems to be a whole Noah's Ark of such appellations. They vary from reference to the smallest insect, as in the case of the gesneriad *Aeschynanthus myrmecophilus* (*myrmecophila, myrmecophilum*, meaning ant-loving), to the largest of creatures, such as *Yucca elephantipes* (*elephantipes*, resembling an elephant's foot).

As with so many plant names, the relationship and resemblance between the common and Latin versions is not always straightforward, logical or even very apparent. Take, for instance, *Fritillaria meleagris*, a very lovely wildflower native to British meadows. Due to the distinctive serpent-like shape of its nodding flower head, this plant is commonly known as the snake's head fritillary, yet the species name refers instead to the decorative patterning of the petals, as *meleagris* (*meleagris, meleagre*) means spotted like a guinea fowl. Several animal terms are used to describe distinctive or unusual markings; for instance, *pardalinus* (*pardalina, pardalinum*), like the spots of a leopard, occurs in the name *Gladiolus pardalinus*. *Zebrinus* (*zebrina, zebrinum*) refers to the stripes of a zebra and describes the highly patterned leaves of *Miscanthus sinensis* 'Zebrinus', the zebra grass.

As a point of interest, the Latin term *colubrinus* (*colubrina, colubrinum*) means like a snake, as in the tree *Anadenanthera colubrina*, while *columbarius* (*columbaria, columbarium*), is dove-like, as in the name for the pincushion flower *Scabiosa columbaria*, sometimes called pigeon's scabious. The genus name for larkspur, *Delphinium*, is from the Greek for dolphin and stems from an old notion that the flower and the marine mammal bear some resemblance. It is not easy to fathom the link between the popular bulbous flower *Hippeastrum puniceum* (syn. *Amaryllis equestris*), and a horse, yet *equestris* (*equestris, equestre*) means relating to horses or horse riders, while *Hippeastrum* comes from the Greek *hippeos*, a mounted man, and *astron*, a star. Rather more straightforward is *puniceus*

**Rosa canina,
Dog rose**

This simple and lovely rose is commonly found in hedgerows.

Hippeastrum puniceum,
Barbados lily

Many perplexed gardeners may ask why the genus name
in this instance relates to horses!

Dracocephalum thymiflorum is a herb with the
common name thyme-leaf dragonhead (*dracocepha-
lus, dracocephala, dracocephalum*, meaning dragon-
headed). A plant with the term *dracunculus* in its
name makes for a less daunting encounter; this refers
to small dragons only, as in *Dracunculus vulgaris*,
the dramatic-looking dragon arum or snake lily
(*dracunculus, dracuncula, dracunculum*).

(*punicea, puniceum*), as it simply means reddish-purple
in colour. Plants as diverse as orchids and broad beans
also have names relating to horses, such as the
commonly termed horseshoe orchid, *Ophrys ferrum-
equinum*, and the horse bean *Vicia faba* var. *equina*
(*equinus, equina, equinum*, of horses).

Names that allude to animals can act as a warning,
as in the makrut lime, *Citrus hystrix*, the trunks and
branches of which have sharp spines about 4 cm
(1.5 in) long. *Hystrix* means bristly or like a porcupine. Similarly, *Dianthus erinaceus* is named after the
hedgehog, as the plant's foliage forms a tight prickly
mound (*erinaceus, erinacea, erinaceum*, meaning like a
hedgehog). Mythological creatures can also be found
in plant names, including several species named after
dragons. One of the most striking is the dragon tree
Dracaena draco, a native of the Spanish Canary
Islands; *draco* means dragon.

Fritillaria meleagris,
Snake's head fritillary

Meleagris describes the decorative petals that are
patterned like the spotted feathers of the guinea fowl.

T

tabularis *tab-yoo-LAH-ris*
tabularis, tabulare
tabuliformis *tab-yoo-lee-FORM-is*
tabuliformis, tabuliforme
Flat, as in *Blechnum tabulare*

tagliabuanus *tag-lee-ah-boo-AH-nus*
tagliabuana, tagliabuanum
Commemorates Alberto and Carlo Tagliabue, 19th-century Italian
nurserymen, as in *Campsis × tagliabuana*

taiwanensis *tai-wan-EN-sis*
taiwanensis, taiwanense
From Taiwan, as in *Chamaecyparis taiwanensis*

takesimanus *tak-ess-ih-MAH-nus*
takesimana, takesimanum
Connected with the Liancourt Rocks (Takeshima in Japanese), as in
Campanula takesimana

taliensis *tal-ee-EN-sis*
taliensis, taliense
From the Tali Range, Yunnan, China, as in *Lobelia taliensis*

Prumnopitys taxifolia,
black pine

tanacetifolius *tan-uh-kee-tih-FOH-lee-us*
tanacetifolia, tanacetifolium
With leaves like tansy (*Tanacetum*), as in *Phacelia tanacetifolia*

tangelo *TAN-jel-oh*
A hybrid of tangerine (*Citrus reticula*) and pomelo (*C. maxima*), as
in *Citrus × tangelo*

tanguticus *tan-GOO-tih-kus*
tangutica, tanguticum
Connected with the Tangut region of Tibet, as in *Daphne tangutica*

tardiflorus *tar-dee-FLOR-us*
tardiflora, tardiflorum
Flowering late in the season, as in *Cotoneaster tardiflorus*

tardus *TAR-dus*
tarda, tardum
Late, as in *Tulipa tarda*

tasmanicus *tas-MAN-ih-kus*
tasmanica, tasmanicum
Connected with Tasmania, Australia, as in *Dianella tasmanica*

tataricus *tat-TAR-ih-kus*
tatarica, tataricum
Connected with the historical region of Tartary (now the Crimea),
as in *Lonicera tatarica*

tatsienensis *tat-see-en-EN-sis*
tatsienensis, tatsienense
From Tatsienlu, China, as in *Delphinium tatsienense*

tauricus *TAW-ih-kus*
taurica, tauricum
Connected with Taurica (now Crimea), as in *Onosma taurica*

taxifolius *taks-ih-FOH-lee-us*
taxifolia, taxifolium
With leaves like yew (*Taxus*), as in *Prumnopitys taxifolia*

tazetta *taz-ET-tuh*
Little cup, as in *Narcissus tazetta*

tectorum *tek-TOR-um*
Of house roofs, as in *Sempervivum tectorum*

temulentus *tem-yoo-LEN-tus*
temulenta, temulentum
Inebriated, as in *Lolium temulentum*

tenax *TEN-aks*
Tough; matted, as in *Phormium tenax*

tenebrosus *teh-neh-BROH-sus*
tenebrosa, tenebrosum
Connected with dark and shady places, as in *Catasetum tenebrosum*

tenellus *ten-ELL-us*
tenella, tenellum
Tender; delicate, as in *Prunus tenella*

tener *TEN-er*
tenera, tenerum
Slender; soft, as in *Adiantum tenerum*

tentaculatus *ten-tak-yoo-LAH-tus*
tentaculata, tentaculatum
With tentacles, as in *Nepenthes tentaculata*

tenuis *TEN-yoo-is*
tenuis, tenue
Slender; thin, as in *Bupleurum tenue*

tenuicaulis *ten-yoo-ee-KAW-lis*
tenuicaulis, tenuicaule
With slender stems, as in *Dahlia tenuicaulis*

tenuiflorus *ten-yoo-ee-FLOR-us*
tenuiflora, tenuiflorum
With slender flowers, as in *Muscari tenuiflorum*

tenuifolius *ten-yoo-ih-FOH-lee-us*
tenuifolia, tenuifolium
With slender leaves, as in *Pittosporum tenuifolium*

tenuissimus *ten-yoo-ISS-ih-mus*
tenuissima, tenuissimum
Very slender; thin, as in *Stipa tenuissima*

tequilana *te-kee-lee-AH-nuh*
Connected with Tequila (Jalisco), Mexico, as in *Agave tequilana*

terebinthifolius *ter-ee-binth-ih-FOH-lee-us*
terebinthifolia, terebinthifolium
With leaves that smell of turpentine, as in *Schinus terebinthifolius*

teres *TER-es*
With a cylindrical form, as in *Vanda teres*

terminalis *term-in-AH-lis*
terminalis, terminale
Ending, as in *Erica terminalis*

ternatus *ter-NAH-tus*
ternata, ternatum
With clusters of three, as in *Choisya ternata*

terrestris *ter-RES-tris*
terrestris, terrestre
From the ground; growing in the ground, as in *Lysimachia terrestris*

tessellatus *tess-ell-AH-tus*
tessellata, tessellatum
Chequered, as in *Indocalamus tessellatus*

testaceus *test-AY-see-us*
testacea, testaceum
Brick-coloured, as in *Lilium × testaceum*

LATIN IN ACTION

This lovely peony, with its fine and elegant leaves, was one of the plants described by Linnaeus in the 1750s. Originating from Russia, it is happiest growing in moist, well-drained woodland sites.

Paeonia tenuifolia
fern leaf peony

testicularis *tes-tik-yoo-LAY-ris*
testicularis, testiculare
Shaped like testicles, as in *Argyroderma testiculare*

testudinarius *tes-tuh-din-AIR-ee-us*
testudinaria, testudinarium
Shaped like a tortoise shell, as in *Durio testudinarius*

tetra-
Used in compound words to denote four

tetragonus *tet-ra-GON-us*
tetragona, tetragonum
With four angles, as in *Nymphaea tetragona*

tetrandrus *tet-RAN-drus*
tetrandra, tetrandrum
With four anthers, as in *Tamarix tetrandra*

tetraphyllus *tet-ruh-FIL-us*
tetraphylla, tetraphyllum
With four leaves, as in *Peperomia tetraphylla*

tetrapterus *tet-rap-TER-us*
tetraptera, tetrapterum
With four wings, as in *Sophora tetraptera*

texanus *tek-SAH-nus*
texana, texanum
texensis *tek-SEN-sis*
texensis, texense
Of or from Texas, USA, as in *Echinocactus texensis*

textilis *teks-TIL-is*
textilis, textile
Relating to weaving, as in *Bambusa textilis*

thalictroides *thal-ik-TROY-deez*
Resembling meadow rue (*Thalictrum*), as in
Anemonella thalictroides

thibetanus *ti-bet-AH-nus*
thibetana, thibetanum
thibeticus *ti-BET-ih-kus*
thibetica, thibeticum
Connected with Tibet, as in *Rubus thibetanus*

thomsonii *tom-SON-ee-eye*
Named after Dr Thomas Thomson, 19th-century Scottish naturalist
and Superintendent of the Calcutta Botanic Garden, India, as in
Clerodendrum thomsoniae

LATIN IN ACTION

This is an illustration of *Chrysanthemum thunbergii*, which was collected at the Cape by Swedish doctor and botanist Carl Peter Thunberg (see p. 72) during his South African plant-hunting expedition in the 1770s. The genus *Thunbergia* is named in his honour and includes the popular brightly coloured climber *Thunbergia alata*, black-eyed Susan (*alata* meaning winged). In its native Tropical Africa, it is classed as a perennial, but in cooler climates it is grown as an annual. Thunberg later went to Japan and collected many specimens there, including *Berberis thunbergii* 'Atropurpurea', also known as Thunberg's barberry (a plant with the term *atropurpurea* as part of its name is dark purple); few plants make a more vividly coloured hedge. Another of Thunberg's introductions from Japan is the lovely and delicate *Fritillaria thunbergii*.

Chrysanthemum thunbergii

thunbergii *thun-BERG-ee-eye*
Named after Carl Peter Thunberg (1743–1828), Swedish botanist, as in *Spiraea thunbergii*

thymifolius *ty-mih-FOH-lee-us*
thymifolia, thymifolium
With leaves like thyme (*Thymus*), as in *Lythrum thymifolium*

thymoides *ty-MOY-deez*
Resembling thyme (*Thymus*), as in *Eriogonum thymoides*

thyrsiflorus *thur-see-FLOR-us*
thyrsiflora, thyrsiflorum
With thyrse-like flower clusters, a central spike with side branches also bearing flower clusters, as in *Ceanothus thyrsiflorus*

thyrsoideus *thurs-OY-dee-us*
thyrsoidea, thyrsoideum
thyrsoides *thurs-OY-deez*
Like a Bacchic staff, as in *Ornithogalum thyrsoides*

tiarelloides *tee-uh-rell-OY-deez*
Resembling *Tiarella*, as in × *Heucherella tiarelloides*

tibeticus *ti-BET-ih-kus*
tibetica, tibeticum
Connected with Tibet, as in *Roscoea tibetica*

tigrinus *tig-REE-nus*
tigrina, tigrinum
With stripes like the Asiatic tiger or with spots like a jaguar (known as 'tiger' in South America), as in *Faucaria tigrina*

tinctorius *tink-TOR-ee-us*
tinctoria, tinctorium
Used as a dye, as in *Genista tinctoria*

tingitanus *ting-ee-TAH-nus*
tingitana, tingitanum
Connected with Tangiers, as in *Lathyrus tingitanus*

titanus *ti-AH-nus*
titana, titanum
Enormous, as in *Amorphophallus titanum*

tobira *TOH-bir-uh*
From the Japanese name, as in *Pittosporum tobira*

tomentosus *toh-men-TOH-sus*
tomentosa, tomentosum
Very woolly; matted, as in *Paulownia tomentosa*

tommasinianus *toh-mas-see-nee-AH-nus*
tommasiniana, tommasinianum
Named after Muzio Giuseppe Spirito de' Tommasini, 19th-century Italian botanist, as in *Campanula tommasiniana*

torreyanus *tor-ree-AH-nus*
torreyana, torreyanum
Named after Dr John Torrey (1796–1873), American botanist, as in *Pinus torreyana*

tortifolius *tor-tih-FOH-lee-us*
tortifolia, tortifolium
With twisted leaves, as in *Narcissus tortifolius*

tortilis *TOR-til-is*
tortilis, tortile
Twisted, as in *Acacia tortilis*

tortus *TOR-tus*
torta, tortum
Twisted, as in *Masdevallia torta*

tortuosus *tor-tew-OH-sus*
tortuosa, tortuosum
Very twisted, as in *Arisaema tortuosum*

totara *toh-TAR-uh*
From the Maori name for this tree, as in *Podocarpus totara*

tournefortii *toor-ne-FOR-tee-eye*
Named after Joseph Pitton de Tournefort (1656–1708), French botanist, first to define the genus, as in *Crocus tournefortii*

townsendii *town-SEN-dee-eye*
Named after David Townsend (1787–1856), American botanist, as in *Spartina* × *townsendii*

toxicarius *toks-ih-KAH-ree-us*
toxicaria, toxicarium
Poisonous, as in *Antiaris toxicaria*

trachyspermus *trak-ee-SPER-mus*
trachysperma, trachyspermum
With rough seeds, as in *Sauropus trachyspermus*

tragophylla *tra-go-FIL-uh*
Literally goat leaf, as in *Lonicera tragophylla*

transcaucasicus *tranz-kaw-KAS-ih-kus*
transcaucasica, transcaucasicum
Connected with Caucasus, Turkey, as in *Galanthus transcaucasicus*

transitorius *tranz-ee-TAW-ree-us*
transitoria, transitorum
Short-lived, as in *Malus transitoria*

transsilvanicus *tranz-il-VAN-ih-kus*
transsilvanica, transsilvanicum
transsylvanicus
transsylvanica, transsylvanicum
Connected with Romania, as in *Hepatica transsilvanica*

trapeziformis *tra-pez-ih-FOR-mis*
trapeziformis, trapeziforme
With four unequal sides, as in *Adiantum trapeziforme*

traversii *trav-ERZ-ee-eye*
Named after William Travers (1819–1903), New Zealand lawyer
and plant collector, as in *Celmisia traversii*

tremulus *TREM-yoo-lus*
tremula, tremulum
Quivering; trembling, as in *Populus tremula*

Asplenium trichomanes,
maidenhair spleenwort

tri-
Used in compound words to denote three

triacanthos *try-a-KAN-thos*
With three spines, as in *Gleditsia triacanthos*

triandrus *TRY-an-drus*
triandra, triandrum
With three stamens, as in *Narcissus triandrus*

triangularis *try-an-gew-LAH-ris*
triangularis, triangulare
triangulatus *try-an-gew-LAIR-tus*
triangulata, triangulatum
With three angles, as in *Oxalis triangularis*

tricho-
Used in compound words to denote hairy

trichocarpus *try-ko-KAR-pus*
trichocarpa, trichocarpum
With hairy fruit, as in *Rhus trichocarpa*

trichomanes *try-KOH-man-ees*
Relating to a Greek name for fern, as in *Asplenium trichomanes*

trichophyllus *try-koh-FIL-us*
tricophylla, tricophyllum
With hairy leaves, as in *Ranunculus trichophyllus*

trichotomus *try-KOH-toh-mus*
trichotoma, trichotomum
With three branches, as in *Clerodendrum trichotomum*

tricolor *TRY-kull-lur*
With three colours, as in *Tropaeolum tricolor*

tricuspidatus *try-kusp-ee-DAH-tus*
tricuspidata, tricuspidatum
With three points, as in *Parthenocissus tricuspidata*

trifasciata *try-fask-ee-AH-tuh*
Three groups or bundles, as in *Sansevieria trifasciata*

trifidus *TRY-fee-dus*
trifida, trifidum
Cut in three, as in *Carex trifida*

triflorus *TRY-flor-us*
triflora, triflorum
With three flowers, as in *Acer triflorum*

Even if the distinctive flowers of this aquatic plant are not in bloom, in a pond it is easily identified by its three leaves with their waxy-smooth surface. Other *Trifoliata* (three-leaved) plants include the Japanese bitter orange, *Poncirus trifoliata*.

Menyanthes trifoliata, bogbean

trifoliatus *try-foh-lee-AH-tus*
trifoliata, trifoliatum
trifolius *try-FOH-lee-us*
trifolia, trifolium
With three leaves, as in *Gillenia trifoliata*

trifurcatus *try-fur-KAH-tus*
trifurcata, trifurcatum
With three forks, as in *Artemisia trifurcata*

trigonophyllus *try-gon-oh-FIL-us*
trigonophylla, trigonophyllum
With triangular leaves, as in *Acacia trigonophylla*

trilobatus *try-lo-BAH-tus*
trilobata, trilobatum
trilobus *try-LO-bus*
triloba, trilobum
With three lobes, as in *Aristolochia trilobata*

trimestris *try-MES-tris*
trimestris, trimestre
Of three months, as in *Lavatera trimestris*

trinervis *try-NER-vis*
trinervis, trinerve
With three nerves, as in *Coelogyne trinervis*

tripartitus *try-par-TEE-tus*
tripartita, tripartitum
With three parts, as in *Eryngium × tripartitum*

tripetalus *try-PET-uh-lus*
tripetala, tripetalum
With three petals, as in *Moraea tripetala*

triphyllus *try-FIL-us*
triphylla, triphyllum
With three leaves, as in *Penstemon triphyllus*

triplinervis *trip-lin-ner-vis*
triplinervis, triplinerve
With three veins, as in *Anaphalis triplinervis*

tripteris *TRIPT-er-is*
tripterus *TRIPT-er-us*
triptera, tripterum
With three wings, as in *Coreopsis tripteris*

tristis *TRIS-tis*
tristis, triste
Dull; sad, as in *Gladiolus tristis*

triternatus *try-tern-AH-tus*
triternata, triternatum
Literally three threes, referring to leaf shape, as in *Corydalis triternata*

trivialis *tri-vee-AH-lis*
trivialis, triviale
Common; ordinary; usual, as in *Rubus trivialis*

truncatus *trunk-AH-tus*
truncata, truncatum
Cut square, as in *Haworthia truncata*

tsariensis *sar-ee-EN-sis*
tsariensis, tsariense
From Tsari, China, as in *Rhododendron tsariense*

tschonoskii *chon-OSK-ee-eye*
Named after Sugawa Tschonoski (1841–1925), Japanese botanist and plant collector, as in *Malus tschonoskii*

tsussimensis *tsoos-sim-EN-sis*
tsussimensis, tsussimense
From Tsushima Island, between Japan and Korea,
as in *Polystichum tsussimense*

tuberculatus *too-ber-kew-LAH-tus*
tuberculata, tuberculatum
tuberculosus *too-ber-kew-LOH-sus*
tuberculosa, tuberculosum
Covered in lumps, as in *Anthemis tuberculata*

tuberosus *too-ber-OH-sus*
tuberosa, tuberosum
Tuberous, as in *Polianthes tuberosa*

***Poa trivialis*,**
rough-stalked meadow grass

tubiferus *too-BIH-fer-us*
tubifera, tubiferum
tubulosus *too-bul-OH-sus*
tubulosa, tubulosum
Shaped like a tube or pipe, as in *Clematis tubulosa*

tubiflorus *too-bih-FLOR-us*
tubiflora, tubiflorum
With trumpet-shaped flowers, as in *Salvia tubiflora*

tulipiferus *too-lip-IH-fer-us*
tulipifera, tulipiferum
Producing tulips or tulip-like flowers, as in *Liriodendron tulipifera*

tuolumnensis *too-ah-lum-NEN-sis*
tuolumnensis, tuolumnense
From Tuolumne County, California, USA, as in
Erythronium tuolumnense

tupa *TOO-pa*
Local name for *Lobelia tupa*

turbinatus *turb-in-AH-tus*
turbinata, turbinatum
Swirling around, as in *Aesculus turbinata*

turczaninowii *tur-zan-in-NOV-ee-eye*
Named after Nicholai S. Turczaninov (1796–1863), Russian
botanist, as in *Carpinus turczaninowii*

turkestanicus *tur-kay-STAN-ih-kus*
turkestanica, turkestanicum
Connected with Turkestan, as in *Tulipa turkestanica*

tweedyi *TWEE-dee-eye*
Named after Frank Tweedy, 19th-century American topographer,
as in *Lewisia tweedyi*

typhinus *ty-FEE-nus*
typhina, typhinum
Like *Typha* (reedmace), as in *Rhus typhina*

TROPAEOLUM

The brightly coloured climbing and twining annual flowers that gardeners usually refer to as nasturtiums are more properly called *Tropaeolum*. Linnaeus (see p. 132) named the plant after the Greek for trophy, *tropaion*. With his typically poetic imagination, he associated the plant he probably saw scrambling up a rustic garden post with the iconography of a classical trophy, transforming its large circular leaves and brightly coloured flowers into round shields and golden helmets arranged on a column! In a similar spirit, nasturtiums signify patriotism in the language of flowers. Common names include Indian cress and monk's cress.

Several species are available, with flower colours ranging from red and yellow to orange and all shades in between, including mixed colours appearing on the same plant. They are very easy to grow, germinate quickly and will self-seed happily, making them excellent plants for children to grow. As well as being extremely decorative, the plants are also edible. The peppery-tasting leaves can be tossed into salads, as can the flowers. The seeds are sometimes pickled and used like capers and can also be dried and ground for use as a substitute for black pepper. The roots of *Tropaeolum tuberosum* are also edible.

This watercolour shows the distinctive shape of the *Tropaeolum* genus of flowers. Plant in full sun for maximum blooms.

The genus includes herbaceous annual and perennial species. Again, the Latin name often gives a clue to the hardiness or otherwise of a particular species. *T. peregrinum* (syn. *T. canariense*) is the Canary creeper; it is frost-tender and better grown as an annual in cooler climates (*peregrinus, peregrina, peregrinum* meaning exotic or foreign; *canariensis, canariensis, canariense*, meaning from the Spanish Canary Islands).

The properly named *Nasturtium officinale* is the aquatic herb watercress (*officinalis, officinalis, officinale* denotes a useful plant sold in shops; for example, vegetables or culinary and medicinal herbs). Belonging to the *Brassicaceae* family, it gets its name from the Latin for twisted nose, *nasi tortium*, alluding to the plant's strong smell. It is best grown in a running stream, although some gardeners claim success with large containers of still water.

Nasturtium officinale,
watercress

The common name nasturtium properly refers to the leafy vegetable *Nasturtium officinale*.

U

ulicinus *yoo-lih-SEE-nus*
ulicina, ulicinum
Like gorse (*Ulex*), as in *Hakea ulicina*

uliginosus *ew-li-gi-NOH-sus*
uliginosa, uliginosum
From swampy and wet regions, as in *Salvia uliginosa*

ulmaria *ul-MAR-ee-uh*
Like elm (*Ulmus*), as in *Filipendula ulmaria*

ulmifolius *ul-mih-FOH-lee-us*
ulmifolia, ulmifolium
With leaves like elm (*Ulmus*), as in *Rubus ulmifolius*

umbellatus *um-bell-AH-tus*
umbellata, umbellatum
With umbels, as in *Butomus umbellatus*

umbrosus *um-BROH-sus*
umbrosa, umbrosum
Growing in shade, as in *Phlomis umbrosa*

uncinatus *un-sin-NA-tus*
uncinata, uncinatum
With a hooked end, as in *Uncinia uncinata*

undatus *un-DAH-tus*
undata, undatum
undulatus *un-dew-LAH-tus*
undulata, undulatum
Wavy; undulating, as in *Hosta undulata*

unedo *YOO-nee-doe*
Edible but of doubtful taste, from *unum edo*, I eat one, as in *Arbutus unedo*

unguicularis *un-gwee-kew-LAH-ris*
unguicularis, unguiculare
unguiculatus *un-gwee-kew-LAH-tus*
unguiculata, unguiculatum
With claws, as in *Iris unguicularis*

uni-
Used in compound words to denote one

unicolor *YOO-nee-ko-lor*
Of one colour, as in *Lachenalia unicolor*

uniflorus *yoo-nee-FLOR-us*
uniflora, uniflorum
With one flower, as in *Silene uniflora*

unifolius *yoo-nih-FOH-lee-us*
unifolia, unifolium
With one leaf, as in *Allium unifolium*

LATIN IN ACTION

This is an old variety of auricula (*undulata* refers to the wavy edge of the leaves). To display such delicate flowers to their best advantage, serious growers construct auricula theatres – wall-mounted shelves topped with a roof to keep the rain off the highly prized blooms.

Primula auricula
var. **undulata**

unilateralis *yoo-ne-LAT-uh-ra-lis*
unilateralis, unilaterale
One-sided, as in *Penstemon unilateralis*

uplandicus *up-LAN-ih-kus*
uplandica, uplandicum
Connected with Uppland, Sweden, as in *S
ymphytum × uplandicum*

urbanus *ur-BAH-nus*
urbana, urbanum
urbicus *UR-bih-kus*
urbica, urbicum
urbius *UR-bee-us*
urbia, urbium
From towns, as in *Geum urbanum*

urceolatus *ur-kee-oh-LAH-tus*
urceolata, urceolatum
Shaped like an urn, as in *Galax urceolata*

urens *UR-enz*
Stinging; burning, as in *Urtica urens*

urophyllus *ur-oh-FIL-us*
urophylla, urophyllum
With leaves with a tip like a tail, as in *Clematis urophylla*

ursinus *ur-SEE-nus*
ursina, ursinum
Like a bear, as in *Eriogonum ursinum*

urticifolius *ur-tik-ih-FOH-lee-us*
urticifolia, urticifolium
With leaves like nettle (*Urtica*), as in *Agastache urticifolia*

uruguayensis *ur-uh-gway-EN-sis*
uruguayensis, uruguayense
From Uruguay, South America, as in *Gymnocalycium uruguayense*

urumiensis *ur-um-ee-EN-sis*
urumiensis urumiense
From Urmia, Iran, as in *Tulipa urumiensis*

urvilleanus *ur-VIL-ah-nus*
urvilleana, urvilleanum
Named after J.S.C. Dumont d'Urville (1790–1842), French
botanist and explorer, as in *Tibouchina urvilleana*

ussuriensis *oo-soo-ree-EN-sis*
ussuriensis, ussuriense
From the River Ussuri, Asia, as in *Pyrus ussuriensis*

Ribes uva-crispa,
gooseberry

utahensis *yoo-tah-EN-sis*
utahensis, utahense
From Utah, USA, as in *Agave utahensis*

utilis *YOO-tih-lis*
utilis, utile
Useful, as in *Betula utilis*

utriculatus *uh-trik-yoo-LAH-tus*
utriculata, utrculatum
Like a bladder, as in *Alyssoides utriculata*

uva-crispa *OO-vuh-KRIS-puh*
Curled grape, as in *Ribes uva-crispa*

uva-ursi *OO-va UR-see*
Bear's grape, as in *Arctostaphylos uva-ursi*

uvaria *oo-VAR-ee-uh*
Like a bunch of grapes, as in *Kniphofia uvaria*

ANDRÉ MICHAUX

(1746–1802)

FRANÇOIS MICHAUX

(1770–1855)

André Michaux was a French botanist and explorer; he was one of the very first collectors of plants in America. Born at Satory, near Versailles, France, Michaux's father was a modest farmer who taught the young André about agriculture and the principles and practices of growing healthy plants. André also received an early grounding in Latin and Greek. By his mid-20s, he was married, but tragedy soon struck when his new wife died following the birth of their son, François. Leaving the child with relatives, the newly widowed Michaux went to Paris to study botany, his ambition being to travel abroad collecting new plant specimens. His first expedition was to Persia (Iran) and the Middle East, where he successfully collected for three years. After returning to France, he was officially appointed the King's Botanist. It was in this capacity that Michaux was commissioned to travel to North America as head of a botanical expedition charged with the specific task of finding new, quick-growing types of trees with which to restock the country's forests. Decades of conflict had left the nation's forests sadly depleted, as so much timber had been felled to build warships.

Michaux arrived in New York in 1785, accompanied by his young son. The years spent in America were both professionally and personally rewarding. His horticultural expertise enabled him to set up a 30-acre nursery near Hackensack, New Jersey, in which to raise live plants to send back to France. He made the acquaintance of William Bartram (see p. 98), with whom he exchanged plant knowledge and seeds; a great friendship and sympathy developed between the two men. Michaux also visited Benjamin Franklin in Philadelphia and George Washington at Mount Vernon. Moving south, Michaux then established a much larger garden and nursery at Charleston, South Carolina, extending to over 100 acres. He met with President Thomas Jefferson, with whom he discussed the possibility of undertaking an exploratory expedition to establish a route between the Mississippi River to the Pacific Ocean, a journey that was later to be made successfully by Lewis and Clark in 1804 (see p. 54).

Michaux set out on many plant-hunting trips during his time in America and identified a great number of plants. By the Savannah River he found *Shortia*

During his time in America, André Michaux sent numerous plant species back to France.

'[MICHAUX HAD] RISEN FROM SIMPLE FARMER TO
HAVE A NAME AMONG LEARNED MEN.'

The Marquis de Lafayette

galacifolia; in the Carolina mountains he found
Rhododendron catawbiense, and in Tennessee he
observed and named *Magnolia macrophylla* (also known
as *M. michauxiana* in his honour). He journeyed as far
afield as Canada and the Bahamas. His last American
journey included sailing up the Catawba River and
exploring Knoxville, Nashville and Mississippi.

While Michaux was abroad, the social and political
unrest back in France had exploded into full-scale
revolution. This adversely affected his ability to send
seeds and live plants back home. It also resulted in dire
personal finances, as his official salary was stopped.
To add to his troubles, on the voyage home
to France in 1796 his ship was wrecked
and his journals and some seeds were lost,
although miraculously his herbarium,
although damaged, survived. Once back
in Paris he was rewarded with honours and
distinctions, but his salary remained unpaid.
Subsequent plant-collecting expeditions and travels
included journeys to England, Spain, the Canary
Islands, Mauritius and Madagascar. The latter
destination was to be his last, as while there he
contracted tropical fever and died.

Michaux published several books, including
Histoire des Chênes de l'Amérique (*Oaks of North
America*) in 1801, and *Flora Boreali-Americana*
(*Flora of North America*) in 1803. His son François
was also a botanist; today he is best known for his
three-volume *Histoire des Arbres Forestiers de
l'Amérique Septentrionale* (*Sylva of North America*),
1810–13. Michaux's herbarium is still housed in Paris
and contains more than 2,000 species. Among the
plants named in his honour are *Lilium michauxii*,
Rhus michauxii and *Quercus michauxii*.

Camellia 'Panache'

André Michaux not only exported a great number
of plants from the United States to his homeland, but also
introduced many plants grown in Europe into America.
The genus *Camellia* was one of his loveliest imports.
Other introductions include the silk tree, *Albizia
julibrissin*, and *Lagerstroemia indica*, the crepe myrtle.

V

vacciniifolius *vak-sin-ee-FOH-lee-us*
vacciniifolia, vacciniifolium
With leaves like blueberry (*Vaccinium*), as in *Persicaria vacciniifolia*

vaccinioides *vak-sin-ee-OY-deez*
Resembling blueberry (*Vaccinium*), as in *Rhododendron vaccinioides*

vagans *VAG-anz*
Widely distributed, as in *Erica vagans*

vaginalis *vaj-in-AH-lis*
vaginalis, vaginale
vaginatus *vaj-in-AH-tus*
vaginata, vaginatum
With a sheath, as in *Primula vaginata*

Securigera varia
(syn. *Coronilla varia*),
crown vetch

valdivianus *val-div-ee-AH-nus*
valdiviana, valdivianum
Connected with Valdivia, Chile, as in *Ribes valdivianum*

valentinus *val-en-TEE-nus*
valentina, valentinum
Connected with Valencia, Spain, as in *Coronilla valentina*

variabilis *var-ee-AH-bih-lis*
variabilis, variabile
varians *var-ee-anz*
variatus *var-ee-AH-tus*
variata, variatum
Variable, as in *Eupatorium variabile*

varicosus *var-ee-KOH-sus*
varicosa, varicosum
With dilated veins, as in *Oncidium varicosum*

variegatus *var-ee-GAH-tus*
variegata, variegatum
Variegated, as in *Pleioblastus variegatus*

varius *VAH-ree-us*
varia, varium
Diverse, as in *Calamagrostis varia*

vaseyi *VAS-ee-eye*
Named after George Richard Vasey (1822–1893), American plant collector, as in *Rhododendron vaseyi*

vedrariensis *ved-rar-ee-EN-sis*
vedrariensis, vedrariense
From Verrières-le-Buisson, France, and the nurseries of Vilmorin-Andrieux & Cie, as in *Clematis* × *vedrariensis*

vegetus *veg-AH-tus*
vegeta, vegetum
Vigorous, as in *Ulmus* × *vegeta*

veitchianus *veet-chee-AH-nus*
veitchiana, veitchianum
veitchii *veet-chee-EYE*
Named after members of the Veitch family, nurserymen of Exeter and Chelsea, as in *Paeonia veitchii*

VACCINIUM

There are a considerable number of *Vaccinium* species, and they have a rather bewildering array of common names attached to them. The native American blueberry is *Vaccinium corymbosum*; often referred to as the highbush blueberry, it is in wide cultivation. (*Corymbosus, corymbosa, corymbosum* relates to the plant's corymb or flat-topped flower cluster.) What is usually known as the American blueberry is *V. cyanococcus*, which describes its blue-coloured fruits. The Canadian blueberry is *V. myrtilloides* (resembling *Myrtus*, myrtle), the common names of which include sourtop and velvet leaf. Confusingly, *V. myrtillus* is the whortleberry, also known as whinberry.

There are also bilberries such as *V. deliciosum*, the cascade bilberry (*deliciosus, deliciosa, deliciosum*, meaning delicious). One way to distinguish between the various types is the colour of the flesh of the fruit; blueberries have white or light green flesh, and huckleberries and bilberries have red or purple. The former also have plentiful small seeds, while the latter have far fewer and they are also larger. Common names for this group of fruits abound and include bog blueberry, black-heart berry, cowberry, farkleberry, grouseberry, rabbiteye blueberry, sparkleberry, windberry and whortle-berry. We should also not forget the American cranberry, *V. macrocarpon*!

Belonging to the family *Ericaceae*, this group of deciduous and evergreen hardy shrubs and small trees should be grown in sun or semi-shade and planted in moist but well-drained, acidic soil enriched with plenty of organic matter. The deciduous plants have very good autumn colour, but the vacciniums

Vaccinium crassifolium,
creeping blueberry

are grown primarily for their edible fruits. Due to the high levels of antioxidants found in the berries, in recent studies they have become lauded as something of a 'super food'. Other plants with linguistic connections to blueberry include *Ilex vaccinioides* (resembling *Vaccinium*) and *Quercus vaccinifolia* (with leaves like *Vaccinium*).

Vaccinium macrocarpon, the American cranberry, has smaller fruit than many other berries of the same family.

velutinus *vel-oo-TEE-nus*
velutina, velutinum
Like velvet, as in *Musa velutina*

venenosus *ven-ee-NOH-sus*
venenosa, venenosum
Very poisonous, as in *Caralluma venenosa*

venosus *ven-OH-sus*
venosa, venosum
With many veins, as in *Vicia venosa*

ventricosus *ven-tree-KOH-sus*
ventricosa, ventricosum
With a swelling on one side, belly-like, as in
Ensete ventricosum

venustus *ven-NUSS-tus*
venusta, venustum
Handsome, as in *Hosta venusta*

verbascifolius *ver-bask-ih-FOH-lee-us*
verbascifolia, verbascifolium
With leaves like mullein (*Verbascum*), as in
Celmisia verbascifolia

verecundus *ver-ay-KUN-dus*
verecunda, verecundum
Modest, as in *Columnea verecunda*

veris *VER-is*
Relating to spring; flowering in spring, as in *Primula veris*

vernalis *ver-NAH-lis*
vernalis, vernale
Relating to spring; flowering in spring, as in
Pulsatilla vernalis

vernicifluus *ver-nik-IF-loo-us*
verniciflua, vernicifluum
Producing varnish, as in *Rhus verniciflua*

vernicosus *vern-ih-KOH-sus*
vernicosa, vernicosum
Varnished, as in *Hebe vernicosa*

vernus *VER-nus*
verna, vernum
Relating to spring, as in *Leucojum vernum*

Sciadopitys verticillata,
Japanese umbrella pine

verrucosus *ver-oo-KOH-sus*
verrucosa, verrucosum
Covered with warts, as in *Brassia verrucosa*

verruculosus *ver-oo-ko-LOH-sus*
verruculosa, verruculosum
With small warts, as in *Berberis verruculosa*

versicolor *VER-suh-kuh-lor*
With various colours, as in *Oxalis versicolor*

verticillatus *ver-ti-si-LAH-tus*
verticillata, verticillatum
With a whorl or whorls, as in *Sciadopitys verticillata*

verus *VER-us*
vera, verum
True; standard; regular, as in *Aloe vera*

vescus *VES-kus*
vesca, vescum
Thin; feeble, as in *Fragaria vesca*

vesicarius *ves-ee-KAH-ree-us*
vesicaria, vesicarium
vesiculosus *ves-ee-kew-LOH-sus*
vesiculosa, vesiculosum
Like a bladder; with small bladders, as in *Eruca vesicaria*

vespertinus *ves-per-TEE-nus*
vespertina, vespertinum
Relating to the evening; flowering in the evening, as in
Moraea vespertina

vestitus *ves-TEE-tus*
vestita, vestitum
Covered; clothed, as in *Sorbus vestita*

vexans *VEKS-anz*
Vexatious or troublesome in some respect, as in *Sorbus vexans*

vialii *vy-AL-ee-eye*
Named after Paul Vial (1855–1917), as in *Primula vialii*

vialis *vee-AH-lis*
vialis, viale
From the wayside, as in *Calyptocarpus vialis*

viburnifolius *vy-burn-ih-FOH-lee-us*
viburnifolia, viburnifolium
With leaves like *Viburnum*, as in *Ribes viburnifolium*

viburnoides *vy-burn-OY-deez*
Resembling *Viburnum*, as in *Pileostegia viburnoides*

victoriae *vik-TOR-ee-ay*
victoriae-reginae *vik-TOR-ee-ay re-JEE-nay*
Named after Queen Victoria (1819–1901), British monarch, as in
Agave victoriae-reginae

vigilis *VIJ-il-is*
vigilans *VIJ-il-anz*
Vigilant, as in *Diascia vigilis*

villosus *vil-OH-sus*
villosa, villosum
With soft hairs, as in *Photinia villosa*

vilmorinianus *vil-mor-in-ee-AH-nus*
vilmoriniana, vilmorinianum
vilmorinii *vil-mor-IN-ee-eye*
Named after Maurice de Vilmorin (1849–1918),
French nurseryman, as in *Cotoneaster vilmorinianus*

LATIN IN ACTION

Multicoloured tulips, such as in this illustration of an unknown 'Versicolor', are a wonderful example of flowers that have been cultivated to display petals of more than one colour. There are also a great number of pure single-colour tulips as well as variegated species, such as the *Viridiflora* tulips that have petals streaked with green.

Tulipa 'Versicolor'

viminalis *vim-in-AH-lis*
viminalis, viminale
vimineus *vim-IN-ee-us*
viminea, vimineum
With long, slender shoots, as in *Salix viminalis*

viniferus *vih-NIH-fer-us*
vinifera, viniferum
Producing wine, as in *Vitis vinifera*

violaceus *vy-oh-LAH-see-us*
violacea, violaceum
Violet, as in *Hardenbergia violacea*

violescens *vy-oh-LESS-enz*
Turning violet, as in *Phyllostachys violescens*

virens *VEER-enz*
Green, as in *Penstemon virens*

virescens *veer-ES-enz*
Turning green, as in *Carpobrotus virescens*

virgatus *vir-GA-tus*
virgata, virgatum
Twiggy, as in *Panicum virgatum*

virginalis *vir-jin-AH-lis*
virginalis, virginale
virgineus *vir-JIN-ee-us*
virginea, virgineum
White; virginal, as in *Anguloa virginalis*

virginianus *vir-jin-ee-AH-nus*
virginiana, virginianum
virginicus *vir-JIN-ih-kus*
virginica, virginicum
virgineus *vir-JIN-ee-us*
virginea, virgineum
Connected with Virginia, USA, as in
Hamamelis virginiana

viridi-
Used in compound words to denote green

viridis *VEER-ih-dis*
viridis, viride
Green, as in *Trillium viride*

viridescens *vir-ih-DESS-enz*
Turning green, as in *Ferocactus viridescens*

Prunus virginiana,
chokecherry

viridiflorus *vir-id-uh-FLOR-us*
viridiflora, viridiflorum
With green flowers, as in *Lachenalia viridiflora*

viridissimus *vir-id-ISS-ih-mus*
viridissima, viridissimum
Very green, as in *Forsythia viridissima*

viridistriatus *vi-rid-ee-stry-AH-tus*
viridistriata, viridistriatum
With green stripes, as in *Pleioblastus viridistriatus*

viridulus *vir-ID-yoo-lus*
viridula, viridulum
Rather green, as in *Tricyrtis viridula*

viscidus *VIS-kid-us*
viscida, viscidum
Sticky; clammy, as in *Teucrium viscidum*

viscosus *vis-KOH-sus*
viscosa, viscosum
Sticky; clammy, as in *Rhododendron viscosum*

vitaceus *vee-TAY-see-us*
vitacea, vitaceum
Like vine (*Vitis*), as in *Parthenocissus vitacea*

vitellinus *vy-tel-LY-nus*
vitellina, vitellinum
The colour of egg yolk, as in *Encyclia vitellina*

viticella *vy-tee-CHELL-uh*
Small vine, as in *Clematis viticella*

vitifolius *vy-tih-FOH-lee-us*
vitifolia, vitifolium
With leaves like vine (*Vitis*), as in *Abutilon vitifolium*

vitis-idaea *VY-tiss-id-uh-EE-uh*
Vine of Mount Ida, as in *Vaccinium vitis-idaea*

vittatus *vy-TAH-tus*
vittata, vittatum
With lengthwise stripes, as in *Billbergia vittata*

vivax *VY-vaks*
Long-lived, as in *Phyllostachys vivax*

viviparus *vy-VIP-ar-us*
vivipara, viviparum
Producing plantlets; self-propagating, as in *Persicaria vivipara*

volubilis *vol-OO-bil-is*
volubilis, volubile
Twining, as in *Aconitum volubile*

vomitorius *vom-ih-TOR-ee-us*
vomitoria, vomitorium
Emetic, as in *Ilex vomitoria*

vulgaris *vul-GAH-ris*
vulgaris, vulgare
vulgatus *vul-GAIT-us*
vulgata, vulgatum
Common, as in *Aquilegia vulgaris*

Prosthechea vitellina
(syn. *Epidendrum vitellinum*),
yolk-yellow prosthechea

W

wagnerii *wag-ner-EE-eye*
wagneriana *wag-ner-ee-AH-nuh*
wagnerianus *wag-ner-ee-AH-nus*
Named after Warren Wagner (1920–2000), American botanist, as in *Trachycarpus wagnerianus*

wahlenbergii *wah-len-BERG-gee-eye*
Named after Georg (Göran) Wahlenberg (1780–1851), Swedish naturalist, as in *Luzula wahlenbergii*

walkerae *WAL-ker-ah*
walkeri *WAL-ker-ee*
Commemorates various Walkers, including Ernest Pillsbury Walker (1891–1969), American zoologist, as in *Chylismia walkeri*

wallerianus *wall-er-ee-AH-nus*
walleriana, wallerianum
Named after Horace Waller (1833–96), English missionary, as in *Impatiens walleriana*

wallichianus *wal-ik-ee-AH-nus*
wallichiana, wallichianum
Named after Dr Nathaniel Wallich (1786–1854), Danish botanist and plant hunter, as in *Pinus wallichiana*

walteri *WAL-ter-ee*
Named after Thomas Walter, 18th-century American botanist, as in *Cornus walteri*

wardii *WAR-dee-eye*
Named after Frank Kingdon-Ward (1885–1958), English botanist and plant collector, as in *Roscoea wardii*

warscewiczii *vark-zeh-wik-ZEE-eye*
Named after Joseph Warszewicz (1812–1866), Polish orchid collector, as in *Kohleria warscewiczii*

watereri *wat-er-EER-eye*
Named after Waterers Nurseries, Knaphill, England, as in *Laburnum × watereri*

webbianus *web-bee-AH-nus*
webbiana, webbianum
Named after Philip Barker Webb (1793–1854), English botanist and traveller, as in *Rosa webbiana*

weyerianus *wey-er-ee-AH-nus*
weyeriana, weyerianum
Named after William van de Weyer, 20th-century horticulturist, who bred *Buddleja × weyeriana*

wheeleri *WHEE-ler-ee*
Named after George Montague Wheeler (1842–1905), American surveyor, as in *Dasylirion wheeleri*

wherryi *WHER-ee-eye*
Named after Dr Edgar Theodore Wherry (1885–1982), American botanist and geologist, as in *Tiarella wherryi*

whipplei *WHIP-lee-eye*
Named after Lieutenant Amiel Weeks Whipple (1818–63), American surveyor, as in *Yucca whipplei*

wichurana *whi-choo-re-AH-nuh*
Named after Max Ernst Wichura (1817–1866), German botanist, as in *Rosa wichurana*

wightii *WIGHT-ee-eye*
Named after Robert Wight (1796–1872), botanist and Superintendent of Madras Botanic Garden, as in *Rhododendron wightii*

Rhododendron wightii

wildpretii *wild-PRET-ee-eye*
Named after Hermann Josef Wildpret, 19th-century
Swiss botanist, as in *Echium wildpretii*

wilkesianus *wilk-see-AH-nus*
wilkesiana, wilkesianum
Named after Charles Wilkes (1798–1877), American naval
officer and explorer, as in *Acalypha wilkesiana*

williamsii *wil-yams-EE-eye*
Named for various eminent botanists and horticulturists
called Williams, including John Charles Williams, 19th-
century English plant collector, as in *Camellia × williamsii*

willmottianus *wil-mot-ee-AH-nus*
willmottiana, willmottianum
willmottiae *wil-MOT-ee-eye*
Named after Ellen Willmott (1858–1934), English
horticulturist, of Warley Place, Essex, as in *Rosa willmottiae*

wilsoniae *wil-SON-ee-ay*
wilsonii *wil-SON-ee-eye*
Named after Dr Ernest Henry Wilson (1876–1930),
English plant hunter, as in *Spiraea wilsonii*. The epithet
wilsoniae commemorates his wife Helen

wintonensis *win-ton-EN-sis*
wintonensis, wintonense
From Winchester; used especially of Hillier Nurseries,
Hampshire, as in *Halimiocistus × wintonensis*

wisleyensis *wis-lee-EN-sis*
wisleyensis, wisleyense
Named after RHS Garden Wisley, Surrey, England, as in
Gaultheria × wisleyensis

wittrockianus *wit-rok-ee-AH-nus*
wittrockiana, wittrockianum
Named after Professor Veit Brecher Wittrock (1839–1914),
Swedish botanist, as in *Viola × wittrockiana*

woodsii *WOODS-ee-eye*
Named after Joseph Woods, 19th-century English botanist
and rose expert, as in *Rosa woodsii*

woodwardii *wood-WARD-ee-eye*
Named after Thomas Jenkinson Woodward, (c. 1742–
1820), British botanist, as in *Primula woodwardii*

woronowii *wor-on-OV-ee-eye*
Named after Georg Woronow (1874–1931), Russian
botanist and plant collector, as in *Galanthus woronowii*

LATIN IN ACTION

By crossing *Camellia saluenensis* with *C. japonica*,
John Williams produced a group of hybrid camellias
that have become firm favourites with many gardeners as they are tough, easy to grow and have beautiful
blooms. There are several cultivars, with flower
colours ranging from white to pink and rose-purple.

Camellia × williamsii

wrightii *RITE-ee-eye*
Named after Charles Wright, 19th-century American botanist
and plant collector, as in *Viburnum wrightii*

wulfenianus *wulf-en-ee-AH-nus*
wulfeniana, wulfenianum
wulfenii *wulf-EN-ee-eye*
Named after Franz Xaver, Freiherr von Wulfen (1728–1805),
Austrian botanist and naturalist, as in *Androsace wulfeniana*

X

xanth-
Used in compound words to denote yellow

xanthinus *zan-TEE-nus*
xanthina, xanthinum
Yellow, as in *Rosa xanthina*

xanthocarpus *zan-tho-KAR-pus*
xanthocarpa, xanthocarpum
With yellow fruits, as in *Rubus xanthocarpus*

xantholeucus *zan-THO-luh-cus*
xantholeuca, xantholeucum
Yellow-white, as in *Sobralia xantholeuca*

Rosa xanthina,
canary bird rose

Y

yakushimanus *ya-koo-shim-MAH-nus*
yakushimana, yakushimanum
Connected with Yakushima Island, Japan, as in
Rhododendron yakushimanum

yedoensis *YED-oh-en-sis*
yedoensis, yedoense
yesoensis
yesoensis, yesoense
yezoensis
yezoensis, yezoense
From Tokyo, Japan, as in *Prunus × yedoensis*

yuccifolius *yuk-kih-FOH-lee-us*
yuccifolia, yuccifolium
With leaves like *Yucca*, as in *Eryngium yuccifolium*

yuccoides *yuk-KOY-deez*
Resembling *Yucca*, as in *Beschorneria yuccoides*

yunnanensis *yoo-nan-EN-sis*
yunnanensis, yunnanense
From Yunnan, China, as in *Magnolia yunnanensis*

Z

zabelianus *zah-bel-ee-AH-nus*
zabeliana, zabelianum
Named after Hermann Zabel, 19th-century German
dendrologist, as in *Berberis zabeliana*

zambesiacus *zam-bes-ee-AH-kus*
zambesiaca, zambesiacum
Connected with the Zambezi River, Africa, as in
Eucomis zambesiaca

zebrinus *zeb-REE-nus*
zebrina, zebrinum
With stripes like a zebra, as in *Tradescantia zebrina*

zeyheri *ZAY-AIR-eye*
Named after Karl Ludwig Philipp Zeyher (1799–1859),
German botanist and plant collector, as in *Philadelphus zeyheri*

zeylanicus *zey-LAN-ih-kus*
zeylanica, zeylanicum
From Ceylon (Sri Lanka), as in *Pancratium zeylanicum*

zibethinus *zy-beth-EE-nus*
zebethina, zebethinum
Smelling foul, like a civet cat, as in *Durio zibethinus*

zonalis *zo-NAH-lis*
zonalis, zonale
zonatus *zo-NAH-tus*
zonata, zonatum
With bands, often coloured, as in
Cryptanthus zonatus

LATIN IN ACTION

Zeylanicus tells us a plant is associated with Sri Lanka
(formerly Ceylon). The lovely bulbous plants of
the genus *Crinum* can be found in warm to tropical
regions around the world, including Brazil, several
parts of Africa, India, China and the Seychelles.
Tender in cooler regions, they have trumpet-shaped
flowers with unusual red and white stripes, hence
their descriptive common name. Other plants whose
names indicate their connections with Ceylon include
the bulbous perennial herb *Pancratium zeylanicum*,
known as the rain flower, and the ornamental tree
Cinnamomum zeylanicum (now a syn. of *C. vera*), the
Ceylon cinnamon.

Crinum zeylanicum,
milk and wine lily

Glossary

Anther
The part of the stamen containing pollen.

Awns
The stiff bristly attachments to the bracts of some grass flowers.

Axil
The point between a leaf and a stem from which buds emerge.

Barb
A hook or sharp bristle.

Beard
A growth of hairs, as on the petals of an iris.

Bract
A modified leaf growing at the base of a flower or inflorescence; sometimes brightly coloured.

Calyx
The outer part of a flower that comprises the sepals and protects the bud.

Corolla
Collectively, the petals of a flower.

Corymb
A flat-topped cluster of flowers.

Ear
The grain-bearing part of a cereal plant.

Follicle
A dry, single-chambered fruit that splits along one side to release its seeds.

Frond
The feathery leaf of a fern or palm.

Keel
Structure (usually a petal) resembling the keel of a boat.

Lobe
A rounded projection, generally of a leaf, petal, bract or stipule. A rounded projection, generally of a leaf, petal, bract or stipule.

Node
A stem joint from which the leaves grow.

Ovate
A leaf, bract or petal shaped like an egg, with a broad base.

Palmate
Shaped like an open hand.

Pedicel
The stalk of an individual flower.

Peduncle
The stalk of an inflorescence.

Pericarp
The fruit wall.

Pinnate
Describing leaflets arranged in opposite pairs.

Pistil
The female portion of a flower.

Raceme
An inflorescence in which the youngest flowers are at the apex.

Scale
One of a number of overlapping plates, sometimes referring to the layers of a bulb.

Scape
A leafless flowering stem.

Sepal
The leaf-like structures that protect the flower in bud.

Spathe
A large bract surrounding the flower spike, especially in plants of the family *Araceae*.

Spur
A short lateral branch of a tree or a tubular outgrowth of a flower, often containing nectar.

Stamen
The male organ of a flower.

Stigma
The apex of the female portion of the flower upon which pollen is deposited.

Stipule
The leaf-like outgrowths at the base of a leaf stalk.

Style
The stalk linking the ovary to the stigma of a female flower.

Umbel
A flower cluster from which all the flower stalks arise from a single point.

Wing
A lateral part or projection.

BIBLIOGRAPHY

Burke, Anna L. (Editor). *The Language of Flowers*. London: Hugh Evelyn, 1973.

Brickell, C. (Editor). *The Royal Horticultural Society A–Z Encyclopedia of Garden Plants*. London: Dorling Kindersley, 2008.

Brickell, C. (Editor). *International Code of Nomenclature for Cultivated Plants*. Leuven: ISHS, 2009.

Cubey, J. (Editor). *RHS Plant Finder 2011-2012*. London: Royal Horticultural Society, 2011.

Fara, Patricia. *Botany and Empire: The Story of Carl Linnaeus and Joseph Banks*. London: Icon Books, 2004.

Fry, Carolyn. *The Plant Hunters*. London: Andre Deutsch, 2009.

Gledhill, D. *The Names of Plants*. Cambridge University Press, 2008

Hay, Roy (Editor). *Reader's Digest Encyclopedia of Garden Plants and Flowers*. London: Reader's Digest, 1985.

Hillier, J. and A. Coombes (Editors). *The Hillier Manual of Trees and Shrubs*. Newton Abbot: David & Charles, 2007.

Johnson, A.T. and H.A. Smith. *Plant Names Simplified*. Ipswich: Old Pond Publishing, 2008.

Neal, Bill. *Gardener's Latin*. New York: Workman Publishing, 1993.

Page, Martin (foreword). *Name That Plant An Illustrated Guide to Plant and Botanical Latin Names*. Cambridge: Worth Press, 2008.

Payne, Michelle. *Marianne North, A Very Intrepid Painter*. London: Kew Publishing, 2011.

Smith, A.W.. *A Gardener's Handbook of Plant Names: Their Meaning and Origins*. New York: Dover Publications, 1997.

Stearn, William T. *Botanical Latin*. Portland: Timber Press, 2004.

Stearn, William T. *Stearn's Dictionary of Plant Names for Gardeners: A Handbook on the Origin and Meaning of the Botanical Names of Some Cultivated Plants*. London: Cassell, 1996.

Wells, Diana. *100 Flowers and How They Got Their Names*. New York: Workman Publishing, 1997.

Websites

Arnold Arboretum, Harvard University
www.arboretum.harvard.edu

Backyard Gardener
www.backyardgardener.com

Chelsea Physic Garden, London
www.chelseaphysicgarden.co.uk

Dave's Garden
www.davesgarden.com

Explorers' Garden
www.explorersgarden.com

Hortus Botanicus, Amsterdam
www.dehortus.nl

International Plant Names Index
www.ipni.org

Plants Database, US Department of Agriculture
www.plants.usda.gov

Plant Explorers
www.plantexplorers.com

Royal Botanic Gardens, Kew
www.kew.org

Royal Horticultural Society
www.rhs.org.uk

IMAGE CREDITS